Instructor's Manual
to accompany

Technical Mathematics
and
Technical Mathematics with Calculus

Second Edition

Dale Ewen
Joan S. Gary
James E. Trefzger

PEARSON
Prentice
Hall

Upper Saddle River, New Jersey
Columbus, Ohio

V036 20 19 18 17 16 15 14 13 12

PEARSON
Prentice
Hall

ISBN 0-13-110285-0

Contents

Answers to Even-Numbered Exercises

CHAPTER 1

Section 1.1

2. 9 4. 2 6. 1 8. 7 10. 1 12. −6 14 −6 16. −13

18. $\frac{3}{16}$ 20. $-3\frac{17}{24}$ 22. −7 24. −4 26. 10 28. −17 30. 3 32. −2

34. −6 36. 0.7 38. $\frac{13}{20}$ 40. $-3\frac{19}{24}$ 42. −15 44. −8 46. 10 48. −13

50. −12 52. 35 54. −2 56. $\frac{2}{7}$ 58. −4 60. $\frac{1}{3}$ 62. $-1\frac{3}{7}$ 64. 240

66. −240 68. $-\frac{3}{5}$ 70. $\frac{1}{2}$ 72. −$180 74. 380 ft

Section 1.2

2. −4 4. 0 6. 0 8. indeterminate 10. indeterminate

12. meaningless 14. $-\frac{1}{2}$ 16. 0, −1 18. 0, −1, 1 20. 2 22. $\frac{1}{2}$

24. 1 26. 4 28. 19 30. 1 32. 10 34. −7 36. −11 38. 15

40. 50 42. 3 44. −14 46. 4 48. $\frac{7}{4}$

Section 1.3

2. 10^9 4. 10^{-10} 6. 10^6 8. 10^{-6} 10. 10^{-6} 12. 10^8 14. 10^{18} 16. 10^{-40}

18. 10^{-30} 20. 6×10^4 22. 1.264×10^{-5} 24. 3.7×10^5 26. 6.7×10^3

28. 1.17×10^1 30. 1.02×10^{-4} 32. 0.0001105 34. 2.96 36. 0.772

38. 31,400,000 40. 0.00318 42. 0.0000000119 44. 1.44×10^{-11}

46. -4.18×10^{-2} 48. 3.19×10^{-13} 50. 1.19×10^{-14} 52. 2.62×10^1

54. 1.33×10^{-8}

Section 1.4

2. 3 4. 2 6. 3 8. 2 10. 3 12. 3 14. 0.1 m

16. 0.001 mi 18. 1000 km 20. 10 V 22. 0.001 km 24. 0.001 g 26. (a) 256 ft

(b) 256 ft 28. (a) all same (b) all same 30. (a) 164.00 km (b) 4.000 km

32. (*a*) $60,\bar{0}00$ V (b) $60,\bar{0}00$ V 34. (a) 300 yd (b) 300 yd 36. (a) all same

(b) 460 kg 38. (a) 0.05 mi (b) 284.0 mi 40. (a) 30,000 V (b) 1,400,000 V

Section 1.5

2. 273 m 4. 2162 g 6. 23.80 mi 8. 290.6 cm 10. 26.61 m

12. 106.4 cm 14. 9 mm or 0.9 cm 16. 61.4 g 18. 0.315 in. 20. 26.3 mm

22. 2520 Ω 24. 15,000 cm^2 26. 55 km^2 28. 6.9×10^{-5} m^3 30. $3\bar{0}0$ V/A

32. 1.1 km/s 34. 290 V^2/Ω 36. 4.2 g/cm^3 38. 640 ft^3 40. 33 m

42. 28.2 hp 44. 970 m^3

Section 1.6

2. Monomial 4. Binomial 6. Trinomial 8. Binomial 10. Trinomial

12. Binomial 14. 3 16. 5 18. 0 20. 2 22. 10 24. 8

26. $-x^5 + x^3 - 8x^2 + 1$; 5 28. $-2x^6 + 8x^5 - 5x^4 - x$; 6 30. $4z^3 - 6z^2 - 3z + 7$; 3

32. $-1 - a + a^2 + a^3$; 3 34. $2 - x + 6x^3 + 7x^4 - 4x^5$; 5 36. $4 - 6y + 7y^2 - 2y^5 + y^8$; 8

38. 5 40. 5 42. 7 44. 10 46. $-3x^2 + 2x + 2$ 48. $12x^2 - 6x - 6$

50. $2x^2 + 2x + 3$ 52. $2x^2 + 7x - 2$ 54. $21x^2 + 11x - 8$ 56. $3x^2 + 2x + 6$

58. 8 60. $-7x + 16$ 62. $-2x^3 + x + 2$ 64. $-9x^2 + 12x - 9$ 66. $x + 9y$

68. $-5x$ 70. -4 72. 13 74. -2592 76. 64 78. 1 80. $-47/4$

82. 256 84. 25

Section 1.7

2. y^9 4. $30y^7$ 6. c^9 8. $1/y^6$ 10. $18x^8$ 12. $\frac{1}{7y^2}$ 14. x^{10} 16. b^{42}

18. $64m^6$ 20. $81c^{24}$ 22. $a^3/8$ 24. x^{10}/y^5 26. 1 28. 7 30. $-32x^{10}$

32. $-s^{10}$ 34. c^{18} 36. a^8b^{12} 38. $25a^4b^8$ 40. $8a^6b^9c^{12}$ 42. $\frac{-27x^{12}}{64y^6}$

44. $\frac{25x^2y^4}{49x^6}$ 46. $\frac{-8}{27x^9}$ 48. 4 50. 7 52. 12 54. 10^3 56. 7 58. -4

60. 10 62. $2\sqrt{3}$ 64. $4\sqrt{5}$ 66. $5\sqrt{3}$ 68. 10 70. 11 72. 22.2

74. 296 76. 0.961 78. 5.66 80. 6.71 82. 3.52×10^{-3} 84. 2.15×10^{7}

Section 1.8

2. $-72x^{10}$ 4. $28x^6y^3$ 6. $-14x^4y^3z^5$ 8. $-30x^5y^6z^7$ 10. $-5ac - 10bc + 15c^2$

12. $-15x^3 - 5x^2 + 35x$ 14. $-12x^5 - 18x^4 + 6x^3$ 16. $-8a^4b^2 + 24a^3b^4$

18. $16a^4b^5 + 24a^4b^3 - 8a^5b^2$ 20. $30x^2 - 23x + 3$ 22. $15x^2 + 11x - 14$ 24. $25x^2 - 81$

26. $15x^2 + 26x + 8$ 28. $6x^2 + 33x - 18$ 30. $18x^2 + 61xy - 7y^2$

32. $28a^2 - 75ab + 50b^2$ 34. $110x^2 + 13x - 6$ 36. $30x^4 + 69x^2 + 36$ 38. $12x^4 + x^2 - 20$

40. $6x^3 - 26x^2 + 12x + 16$ 42. $15x^4 + 43x^3 - 23x^2 - 58x + 14$

44. $2x^2 - 7xy + 3y^2 + 11x + 2y - 21$ 46. $-8x^4 + 8x^3 + 18x^2 - 34x + 20$

48. $6x^4 - 28x^3 + 22x^2 + 24x$ 50. $16x^2 + 24x + 9$ 52. $36x^2 - 48x + 16$ 54. $9x^4 + 24x^2 + 16$

56. $4x^4 + 32x^3 + 81x^2 + 55x - 25$ 58. $64x^3 + 144x^2 + 108x + 27$ 60. $8 - 48x + 96x^2 - 64x^3$

Section 1.9

2. $-6xy^2$ 4. $4a^3b^2$ 6. $\dfrac{8}{7a^3b^2}$ 8. $-\dfrac{8a^2}{3b}$ 10. $\dfrac{6}{x}$ 12. $\dfrac{a^3}{2b^3}$

14. $6x^2 + 3x + 2$ 16. $9x^2 - 11x - 7$ 18. $4x^2 - 5x + 2 - \dfrac{3}{x}$

20. $3x^5 - 2x^2 + 4x - \dfrac{4}{3x^2}$ 22. $b^2 + \dfrac{3b^3}{2a} - 2ab$ 24. $-\dfrac{5q}{2} + \dfrac{4p^2}{q^2} - \dfrac{2}{p}$

26. $-\dfrac{7mn}{3p^2} - \dfrac{5p}{3n} + \dfrac{3}{m}$ 28. $x + 1$ 30. $2x^2 - x + 3 + \dfrac{5}{x+3}$ 32. $x^2 - 3x + 4$

34. $5x^2 - 6x + 2$ 36. $5x^2 + 4x - 8 + \dfrac{20}{5x-4}$ 38. $4x^3 + 5x - 2 - \dfrac{6}{2x-3}$

40. $x^4 - x^3 + x^2 - x + 1$ 42. $x^2 + 2x + 2$ 44. $8a^3 + 4a^2 + 2a + 1 + \dfrac{2}{2a-1}$

Section 1.10

2. 16 4. -7 6. 20 8. 12 10. 4 12. 7 14. 2 16. 8

18. -3 20. $-\dfrac{11}{6}$ 22. -4 24. 5 26. -27 28. 1 30. -0.5 32. $\dfrac{-30}{13}$

34. 30 36. $\dfrac{1}{3}$ 38. 22 40. 18 42. $-\dfrac{7}{22}$ 44. 0 46. 19.7

48. 0.970   50. 3.67   52. 2.79   54. 355   56. 6.46×10^7

Section 1.11

2. $\lambda = \dfrac{v}{f}$ 4. $I = \dfrac{V}{Z}$ 6. $C_3 = C_T - C_1 - C_2 - C_4$ 8. $t = \dfrac{Q}{I}$ 10. $m = \dfrac{E}{c^2}$

12. $A = \dfrac{\rho L}{R}$ 14. $N = \dfrac{2.5(\text{h.p.})}{D^2}$ 16. $g = \dfrac{wa}{f}$ 18. $f = \dfrac{X_L}{2\pi L}$ 20. $V' = \dfrac{V_p}{p}$

22. $C = \dfrac{5}{9}(F - 32)$ or $C = \dfrac{5F - 160}{9}$ 24. $T_1 = \dfrac{V_1 P_1 T_2}{V_2 P_2}$ 26. $I_2 = \dfrac{I_1 r^2_1}{r^2_2}$

28. $\lambda = \dfrac{4l}{2n+1}$ 30. $d = \dfrac{2A - 2ab}{a+c}$ 32. $v_s = \dfrac{f'v - fv - fv_o}{f'}$ 34. $\Delta T = \dfrac{\Delta \rho}{\alpha \rho}$

36. $f = \dfrac{1}{2\pi C X_c}$ 38. $C_T = \dfrac{C_1 C_2}{C_1 + C_2}$ 40. $C_2 = \dfrac{C_1 C_3 C_T}{C_1 C_3 - C_3 C_T - C_1 C_T}$

42. $f = \dfrac{s_o i}{s_o + s_i}$ 44. $R'' = \dfrac{f R'(n-1)}{f(n-1) - R'}$ 46. $m = \pm\sqrt{\dfrac{13.6 n^2}{13.6 - n^2 hf}}$

48. $Z_2 = \dfrac{I_3 Z_3}{I_T - I_3}$ 50. $R_3 = \dfrac{R_1 R_2 - R_1 R_B - R_2 R_B}{R_B}$

Section 1.12

2. 114 cm   4. 14.4 ft.   6. 79 in.   8. 11.40 m   10. 102°C

12. $8.9 \times 10^{11} N\,\text{m}$ 14. $57.0\, A\Omega$ 16. 411 turns   18. $60.1/F\Omega$

20. $-7.8 \times 10^{-8} C$ 22. $26\bar{0}\ N$ 24. 344 m/s   26. 415 Hz   28. 7.89 Ω

30. $6\bar{0}\ \Omega$ 32. $8\bar{0}$ cm   34. 41.7 m   36. 6.00 Ω

Section 1.13

2. 12, 18   4. 36 in.   6. 38 cm, 42 cm, 42 cm   8. \$1800 @ 8%; \$5000@ 10.5%

10. 40 ft × 70 ft   12. (a) 300 mi/h   (b) 240 mi/h   (c) 1200 miles   14. 90 mi

16. 6 kg of 10%; 4 kg of 15%   18. 400 L of 90%; 200 L of 75%   20. 5 L

22. 794 m   24. 20°, 25°, 75°

Section 1.14

2. $\dfrac{20}{3}$ 4. $\dfrac{48}{1}$ 6. $\dfrac{50}{1}$ 8. $\dfrac{864}{1}$ 10. $\dfrac{64}{1}$ 12. $\dfrac{8}{1}$ 14. (a) 40 lb/in^2

(b) 180 kPa   16. $\dfrac{90}{1}$ 18. $\dfrac{16}{3}$ 20. $\dfrac{32}{5}$ 22. 3.5 gal/min   24. $\dfrac{325}{1}$ 26. $\dfrac{12}{5}$

28. 25 30. 288 32. $\frac{1}{3}$ 34. 2 36. $\frac{bc}{a}$ 38. $\frac{pm}{q}$ 40. $\frac{-a^2}{b}$

42. 477 44. 79.6 46. 1.55 48. 212.5 mi 50. (a) 480 km (b) 300 L

(c) 12 km/L 52. $41\frac{1}{4}$ ft^3 54. 2065 56. \$200; \$250 58. 1000 m × 1200 m

60. 45,000 bolts; 20 machines 62. 45 lb 64. 120 lb 66. 1250 ft 68. 1190

70. 21 lb 72. (a) 15 ft^3 (b) 37.5 ft^3 74. (a) 10 m^3 (b) 25 m^3

(c) 40 m^3

Section 1.15

2. Inverse; $k = 20$ 4. Neither 6. Direct; $k = \frac{2}{7}$ 8. $p = \frac{k}{q}$ 10. $m = kn^2$

12. $d = kef^3$ 14. $m = \frac{k\sqrt{n}}{p^3}$ 16. $k = 9$; 315 18. $k = \frac{9}{20}$; $\frac{27}{100}$ 20. $k = 24$; 864

22. $k = 2$; 1536 24. $k = 216$; 3 26. $k = 0.0797$; 1590 lux 28. $k = 52{,}500$ ft^3/in^2; $21\bar{0}0$ ft^3

30. $k = 1.00$ W/VA; $92\bar{0}$ W 32. $k = 1\bar{0}$ hp/in^2; 160 hp 34. 96 rpm

36. 34 cm 38. 108 h

CHAPTER 2

Section 2.1

2. ∠1 and ∠4   4. ∠2 and ∠3   6. ∠5 and ∠6

8. 125°   10. ∠1 and ∠2, or ∠3 and ∠4, or ∠1 and ∠3, or ∠2 and ∠4

12. 40°   14. 140°   16. 130°

18. 50°   20. 50°

Section 2.2

2. 26.4 in.   4. 232 ft   6. (a) 151 cm^2 (b) 65.5 cm

8. (a) 40,100 ft^2 (b) 1004 ft   10. (a) 402 in^2 (b) 109.2 in

12. No, three equal sides would not satisfy the Pythagorean Theorem.

14. Yes, the measure of their corresponding angles are equal.

16. 90° and 54.25°   18. 68.7 mi   20. 4.24 in.

22. 528 cm^2 24. (a) 24.0 cm (b) 34.0 cm

Section 2.3

2. (a) 82.0 cm (b) $42\bar{0}$ cm^2 4. (a) 812 ft (b) 31,800 ft^2

6. (a) 6.60 cm (b) 2.28 cm^2 8. (a) 7830 ft (b) 3,090,000 ft^2

10. 24,400 ft^2 12. 6.12 mi   14. 12 squares

16. (a) 2130 ft^2 (b) 722 ft   18. 5060 cm^2 20. 1950 cm^2

Section 2.4

2. (a) 42.4 ft (b) 143 ft^2 4. (a) 3610 km (b) 1,040,000 km^2

6. (a) 75.4 ft (b) 452 ft^2 8. (a) $11\bar{0}0$ km (b) 96,200 km^2

10. 78.0 cm   12. 14.9 m

14. (a) 66.0° (b) 33.0° (c) 33.0° (d) 99.0°

16. (a) 40.0 cm (b) 20.0 cm (c) 34.6 cm (d) 60.0 cm

18. (a) 20.0° (b) 40.0° (c) 32.8° 20. (a) 2980 ft^2 (b) 216 ft

22. 16.7 in.

Section 2.5

2. (a) 59.9 m^3 (b) 57.9 m^2 (c) 84.8 m^2

4. (a) 29,400 cm^3 (b) 4290 cm^2 (c) 5810 cm^2

6. 29,200 cm^3 8. (a) 1,790,000 cm^3 (b) 71,300 cm^2

10. (a) 125 cm^3 (b) $10\overline{0}$ cm^2 (c) $15\overline{0}$ cm^2

12. (a) 92.2 ft^2 (b) 51.3 ft^3 14. (a) 151,000 ft^3 (b) 17,400 ft^2

16. (a) 7420 ft^2 (b) 6480 ft^2 (c) 123,000 ft^2 (d) 147,000 ft^3

18. 241 m^2 20. (a) $14\overline{0}$ ft^3 (b) 65.0 ft^3

CHAPTER 3

Section 3.1

2.

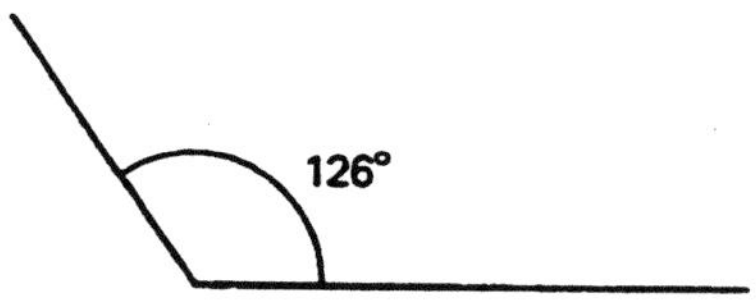

4.

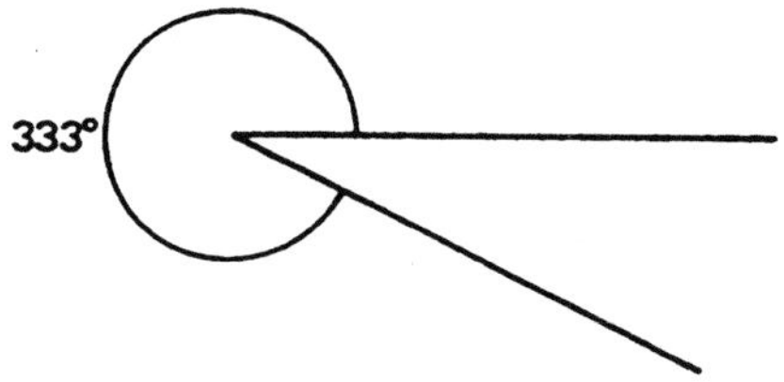

6. 0.1° 8. 0.7833° 10. 40′ 12. 42′ 14. 142.5° 16. 120.1833°

18. 183°30′ 20. 7°24′ 22. 65.46° 24. 135.8025° 26. 35°50′24″

28. 29°36′29″ 30. b 32. b 34. A 36. C 38. A

40. 22.9 ft 42. 59.5 km 44. 2580 mi 46. 527 m 48. 139 mi

	sin A	cos A	tan A	cot A	sec A	csc A
50.	0.590	0.808	0.730	1.37	1.24	1.70
52.	0.489	0.872	0.561	1.78	1.15	2.05
54.	0.629	0.779	0.808	1.24	1.28	1.59

	sin B	cos B	tan B	cot B	sec B	csc B
56.	0.808	0.590	1.37	0.730	1.70	1.24
58.	0.916	0.400	2.29	0.437	2.50	1.09
60	0.537	0.844	0.637	1.57	1.19	1.86
62.	(a) 0.556	0.833	0.667	1.50	1.20	1.80
	(b) 0.556	0.833	0.667	1.50	1.20	1.80

64. tan A = 1.540, cot A = 0.6493, sec A = 1.836, csc A = 1.192

Section 3.2

2. 0.8862 4. 0.2351 6. 19.74 8. 1.014 10. 1.058 12. 0.2849

14. 0.9426 16. 1.7451 18. 3.4022 20. 39.0° 22. 12.1° 24. 53.6°

26. 57.3° 28. 17.0° 30. 10.2° 32. 72.6° 34. 48.6° 36. 6.9°

38. 11.07° 40. 68.64° 42. 37.27° 44. 0.6652 46. 0.3172 48. 0.3607

50. 1.1684 52. 0.7906 54. 8.1887 56. 74°2′38″ 58. 18°1′30″

60. 15°37′43″ 62. 72°35′29″ 64. 12°30′41″ 66. 7°3′22″

Section 3.3

2. A = 46.8°, b = 6.57 m, c = 9.60 m
4. A = 33.7°, B = 56.3°, b = 7.48 in.
6. B = 17.6°, b = 2.92 cm, c = 9.65 cm
8. A = 72.7°, a = 8.97 km, b = 2.80 km
10. A = 7.2°, B = 82.8°, c = 958 yd
12. B = 40.65°, a = 5.539 m, b = 4.755 m
14. A = 58.02°, B = 31.98°, a = 21,670 km
16. A = 65.00°, b = 2332 ft, c = 5517 ft
18. A = 42.43°, B = 47.57°, b = 3.842 mi
20. B = $6\bar{0}$°, b = 26 ft, c = $3\bar{0}$ ft
22. A = 66°, B = 24°, c = 39 ft
24. B = 31°, a = 6200 m, c = 7200 m
26. A = 55°, B = 35°, b = 0.25 mi
28. A = 37°15′57″, B = 52°44′3″, c = 47,180 ft
30. B = 71°, 53′15″, a = 1.751 km, b = 5.353 km
32. A = 23°30′45″, B = 66°29′15″, a = 3373 ft
34. A = 14°52′52″, a = 14.77 km, c = 57.53 km

Section 3.4

2. 6.20 ft
4. 8.2 mi
6. (a) 92.1 ft (b) 17.5 m
8. 184 m
10. $16\bar{0}$ ft, 399 ft
12. 9.40 cm
14. 22.1°, 17.3 ft
16. x = 24.3 m, y = 16.8 m, α = 40.2°, β = 68.8°, θ = 139.8°, θ = 111.2°
18. 318 mm
20. 3.63 cm
22. x = 11.7 m, θ = 59.0°
24. (a) 59.0 in. (b) 25.5° (c) 16.1°
26. 33.6°
28. 13.2 ft
30. 77.95 mm
32. 2.355 cm
34. 33°36′11″; 15.28 in.

CHAPTER 4

Section 4.1

	Function	Domain	Range
2.	Yes	{1, 3, 5}	(2, 3}
4.	No	{0, 2, 5}	{–1, 1, 2, 7}
6.	No	{–3, 3}	{–4, 4}
8.	Yes	Real numbers	Real numbers
10.	Yes	Real numbers	Real numbers where $y \geq -3$
12.	No	Real numbers where $x \geq 4$	Real numbers
14.	Yes	Real numbers where $x \leq \frac{1}{2}$	Real numbers where $y \geq 0$
16.	Yes	Real numbers where $x \geq -5$	Real numbers where $y \leq 16$

18. (a) –4 (b) 20 (c) 32 20. (a) 32 (b) –4 (c) 32 22. (a) –7 (b) –7 (c) –15

24. (a) 4 (b) 6 (c) $\sqrt{11}$ (d) 8 is not in the domain of $g(t)$.

26. (a) $8z^2 - 7z$ (b) $32y^2 - 14y$ (c) $72t^4 - 21t^2$

28. (a) $y^2 - 5y + 10$ (b) $y^4 - y^2 + 4$ (c) $16y^2 + 4y + 4$

30. (a) $2t^2 - 6t + 9$ (b) $-2t - 1$ (c) $t^4 - 8t^3 + 17t^2 - 28t + 20$

(d) $t^2 + 2th + h^2 - 4t - 4h + 4$

32. All real numbers $t \neq -\frac{1}{2}$ 34. All real numbers 36. All real numbers $t > -4$

Section 4.2

2.

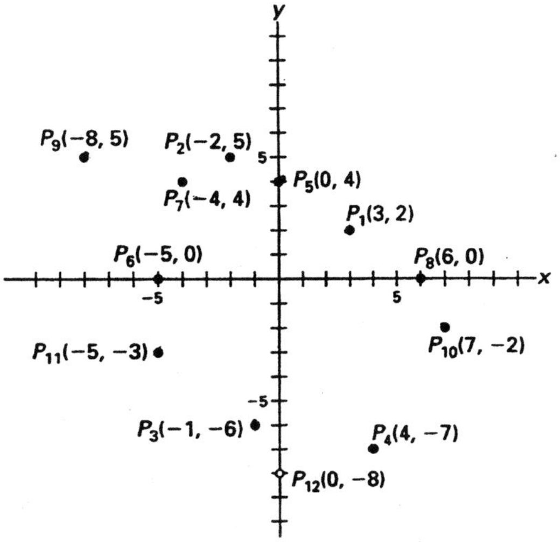

4.

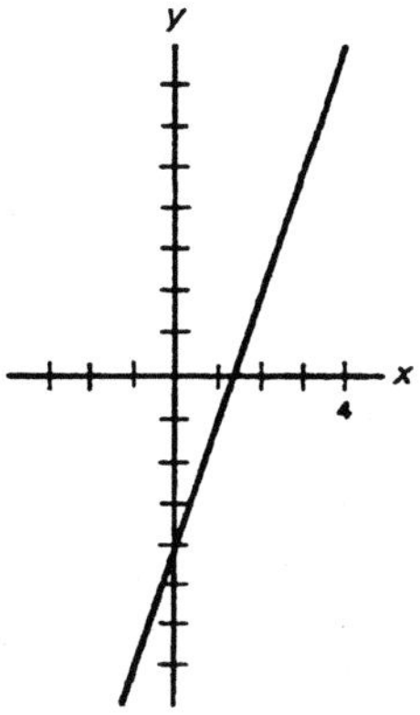

6.

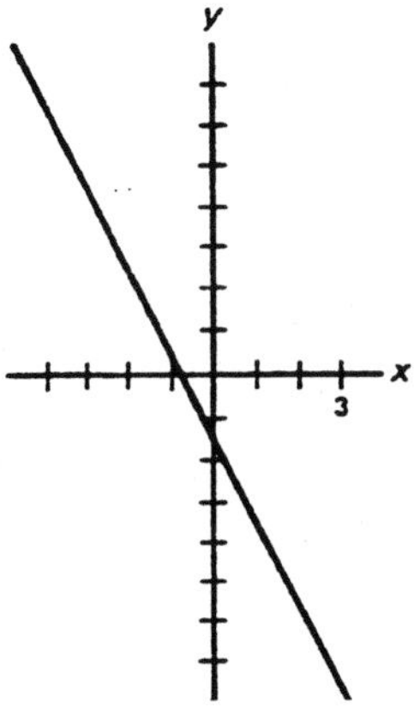

8.

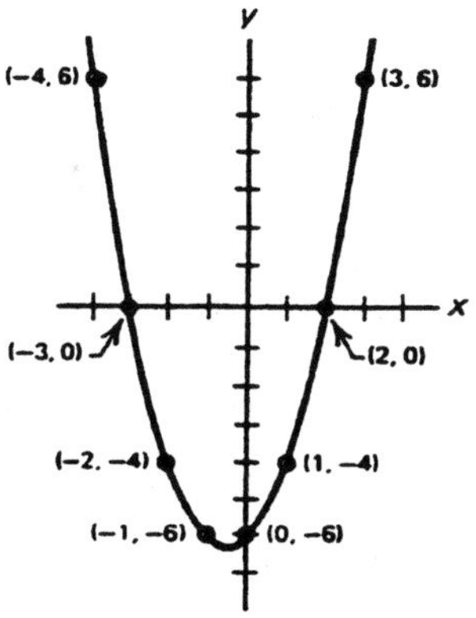

10.

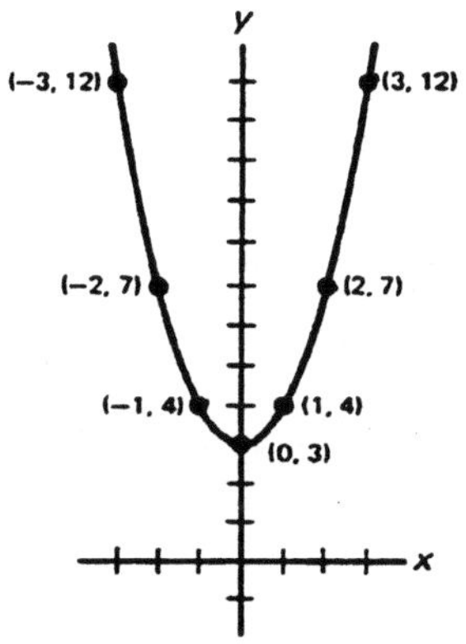

12.

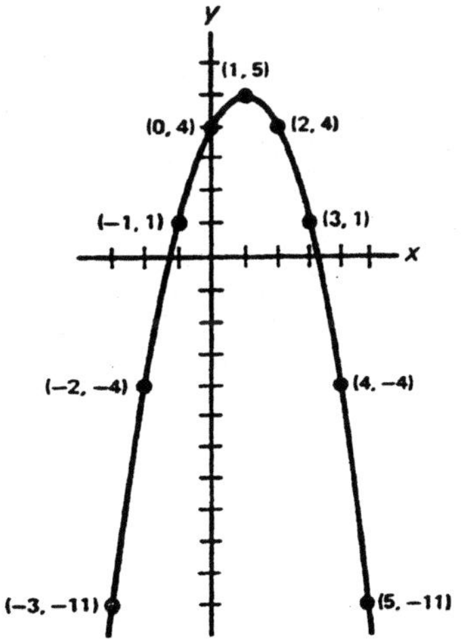

14.

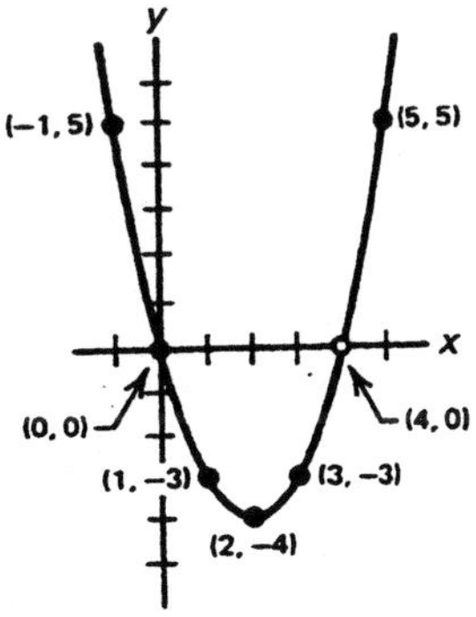

16.

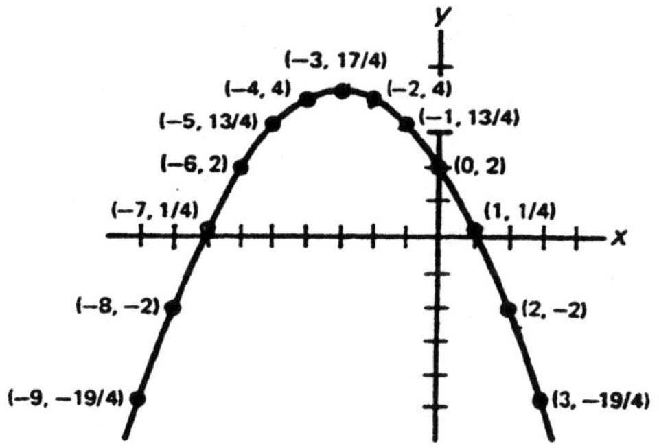

18.

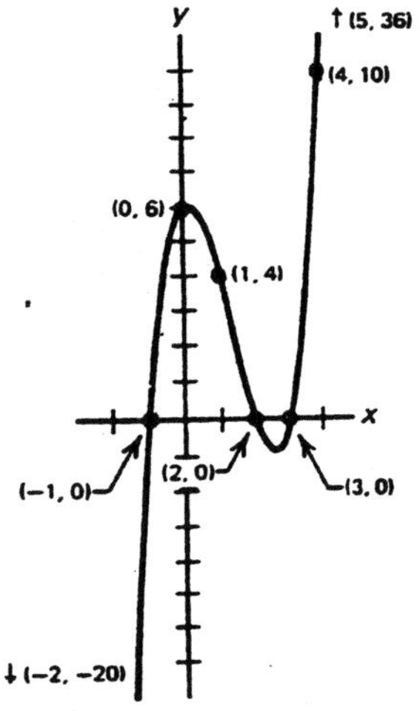

20.

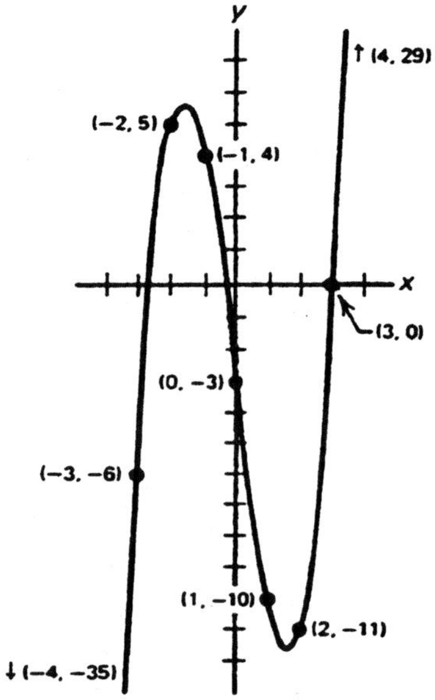

22.

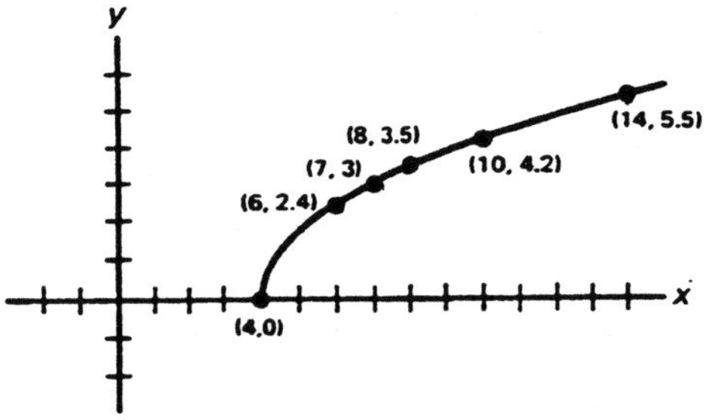

24.

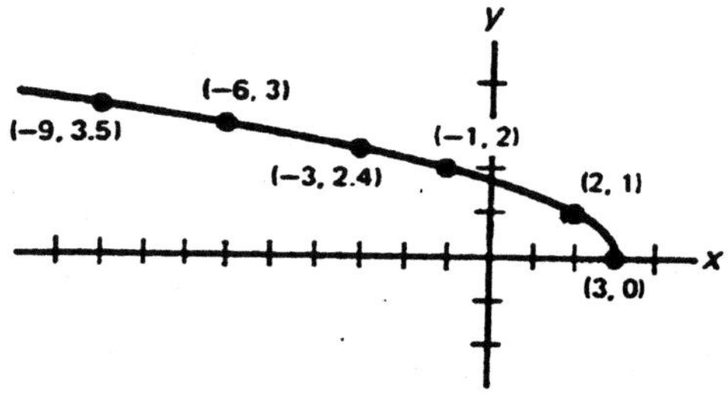

26. −3, 2; −4, 3; 1.3, −2.3

28. No solutions; −1, 1; 1.7, −1.7

30. 3.2, −1.2; 0, 2; 3.6, −1.6

32. 0.4; 0.6, 3.4; 4.7, −0.7

34. 1.1, −7.1; 1.6, −7.6; 0.3, −6.3

36. 3, −1, 2; 3.3, −0.8, 1.5; 3.9

38. 3, −2.6, −0.4; 3.1, −2.4, −0.7; 0, 2.8, −2.8

40. 3.3, −0.8; 3, −0.5; 0, 2.5

42. 0, 4; 3.4, 0.6; 6.5, −2.5

44. 0, 1.7, −1.7; 2; 2.1, −1.5, 0.3

46. 10 s, 16 s, 20 s, 23 s

48. 16.8 ms, 18.5 ms, 22.1 ms

50. −1.8 μs, −2.4 μs

52. A(4, 0), B(2.83, 2.83), C(0, 4), D(−2.83, 2.83), E(−4, 0), F(−2.83, −2.83), G(0, −4), H(2.83, −2.83)

Section 4.3

2. -2 4. $\frac{1}{11}$ 6. Undefined 8. $-\frac{3}{2}$

10.

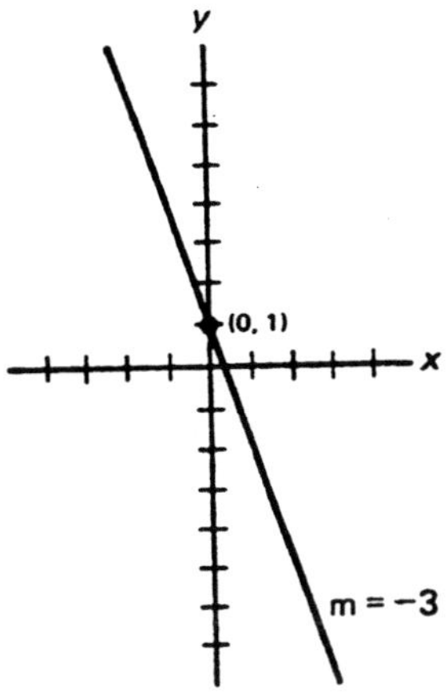

12.

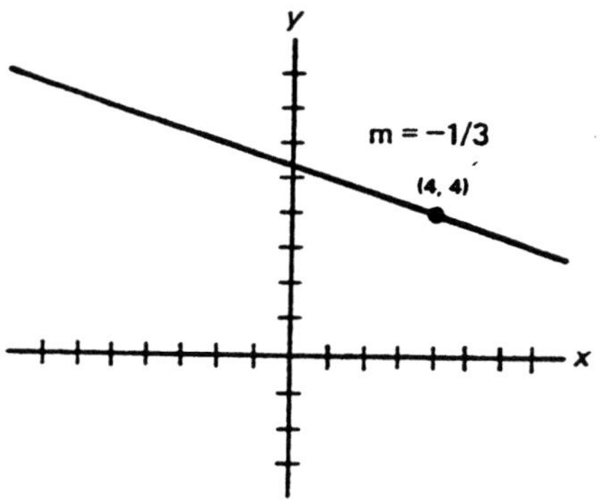

14.

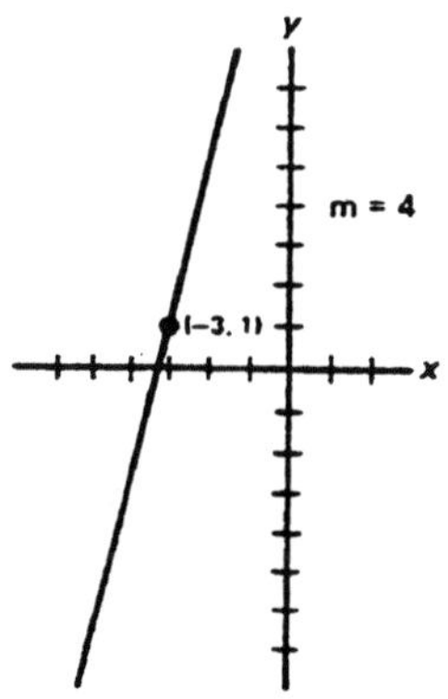

16.

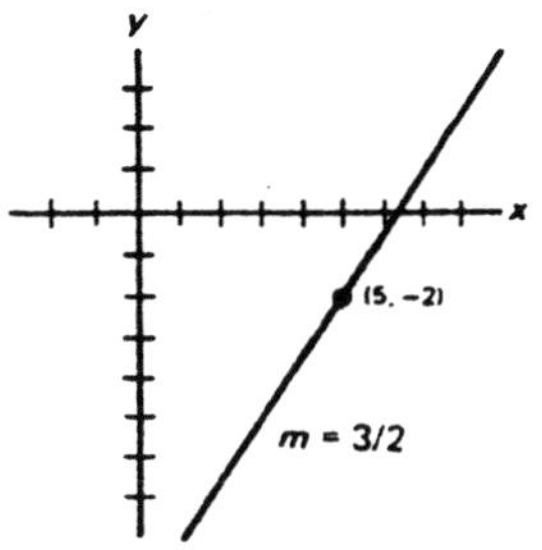

18. $2x - y - 11 = 0$ 20. $3x + 4y + 10 = 0$ .22. $3x + y - 9 = 0$ 24. $x + 3y - 4 = 0$

26. $y = \frac{1}{3}x + 8$ 28. $y = -\frac{3}{4}x - 4$ 30. $y = -2$ 32. $x = 5$ 34. $x = -5$

36. $y = 6$ 38 $m = \frac{2}{3}, b = -3$ 40. $m = \frac{1}{2}, b = 0$

42. No slope, no y-intercept

44.

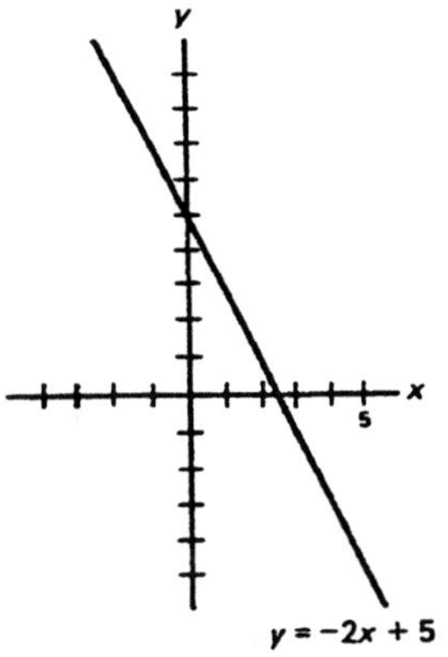

46.

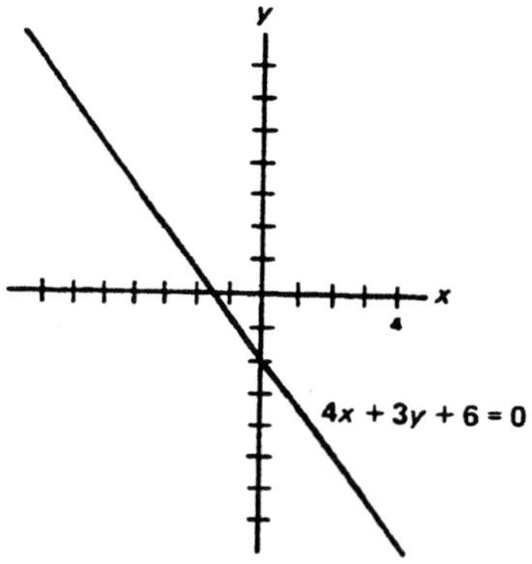

48.

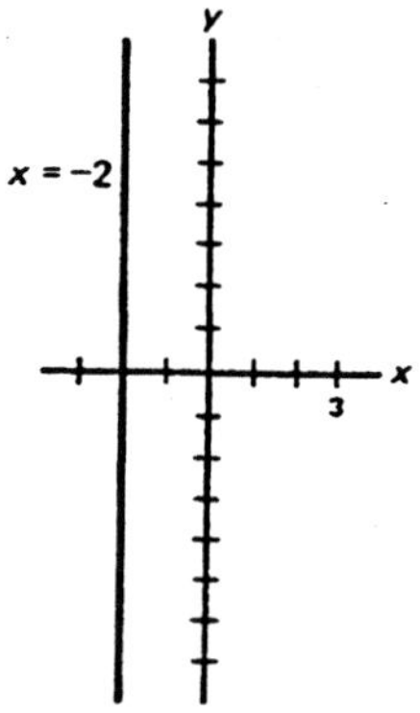

50.

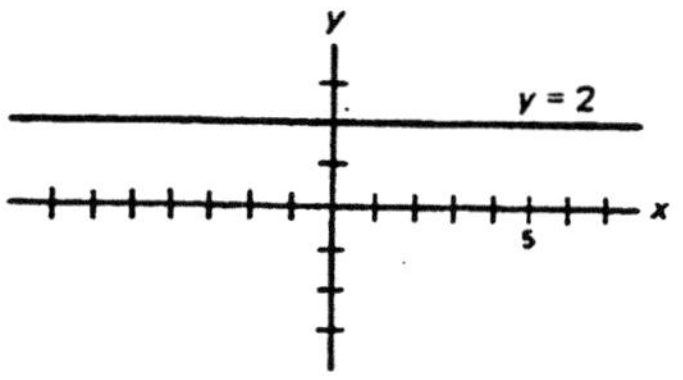

52.

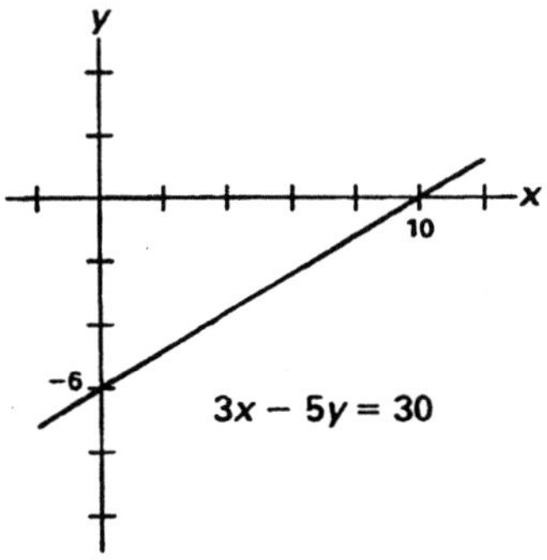

54.

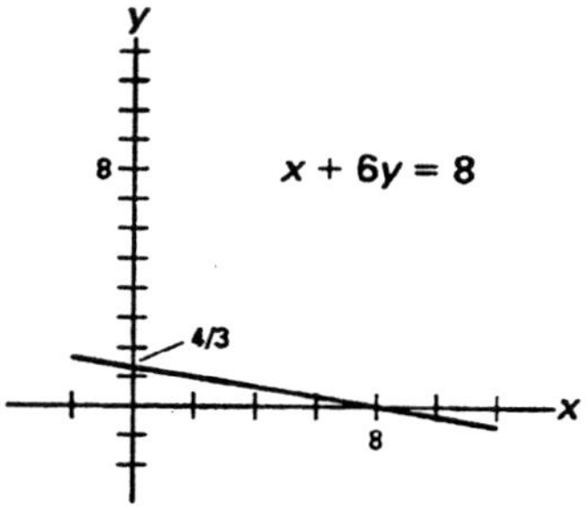

Section 4.4

2. Parallel 4. Perpendicular 6. Neither 8. $2x + 3y + 2 = 0$

10. $2x + 3y + 26 = 0$ 12. $5x - 4y = 0$ 14. $3x - 2y - 4 = 0$ 16. $x = -4$

18. $y = -9$ 20. (a) Yes, $m(AB) = m(CD) = -\dfrac{3}{4}$; $m(AD) = m(BC) = \dfrac{4}{3}$

(b) Yes, $m(AB) = \dfrac{-1}{m(AD)}$

Section 4.5

2. $2\sqrt{13}$ 4. 6 6. 5 8. 4 10. $\left(1, \dfrac{1}{2}\right)$ 12. (3, −3) 14. (−1, 1)

16. (a) $6 + 6\sqrt{2}$ (b) Yes (c) Yes (d) 9 18. (a) $10 + 5\sqrt{2}$ (b) No (c) Yes

(d) NA 20. $5\sqrt{2}$ 22. $5x - 2y = 22$ 24. $5x - 6y = 7$

26. $m(AB) = m(CD) = \frac{1}{3}$, $m(BC) = m(AD) = 3$

28. Use the Pythagorean theorem:

$$(a + c - a)^2 + (b - b)^2 + [(a + c) - (a + c)]^2 + (b + c - b)^2 = (a + c - a)^2 + (b + c - b)^2;$$

$$c^2 + c^2 = 2c^2$$

Since the sum of the squares of the lengths of two sides equals the square of the length of the third, the triangle is a right triangle.

CHAPTER 5

Section 5.1

2. $15x^7 - 20x^5 + 10x^4 - 30x^3 + 35x^2$ 4. $-12a^4b^6 + 30a^3b^6 - 54a^2b^7$

6. $9x^4 - 64y^2z^4$ 8. $64a^4b^2 + 80a^2bc + 25c^2$ 10. $81x^2 - 72xy^3 + 16y^6$

12. $x^2 - x - 72$ 14. $20x^2 - 59x + 42$ 16. $9a^2 + 4b^2 + c^2 - 12ab + 6ac - 4bc$

18. $64x^3 + 240x^2y + 300xy^2 + 125y^3$ 20. $8x^3 - 48x^2y^3 + 96xy^6 - 64y^9$

22. $20x^2 - 44xy + 8y^2$ 24. $4x^2 - 40xy + 100y^2$ 26. $4x^4y^2 + 28x^2yz^3 + 49z^6$

28. $64 - 16a^2b^3 + a^4b^6$ 30. $6x^6y^4z - 10\,x^4y^4z^4 - 12\,x^2y^4z^2 + 8\,x^4y^3z^2$

32. $180a^2 - 51ab - 10b^2$ 34. $-21x^2 + 40x + 21$ 36. $6400 - a^{10}$ 38. $12x^4 + 13x^2 - 35$

40. $25y^2 - 10yz + z^2$ 42. $1 - 2a^2 + a^4$ 44. $4x^4 + \frac{3}{2}x^2y^3 + \frac{1}{8}y^6$

46. $16x^4 - 4x^2y + \frac{1}{4}y^2$ 48. $x^2 - 3x^4 + 3x^6 - x^8$

Section 5.2

2. $5(3x - 4)$ 4. $2x(4x - 6y + 5z)$ 6. $5x(2x + 5)$ 8. $2x^2(1 - 50\,x + 2x^2)$

10. $-16x\,(x + 2)(x + 1)$ 12. $3x^3y(5x^4y^2 - 2x^3y + 4z)$

14. $(x + 8)(x - 8)$ 16. $(7x + 10y)(7x - 10y)$ 18. $5(2x + 3)(2x - 3)$

20. $4(4x + y)(4x - y)$ 22. $(7 + a^2)(7 - a^2)$ 24. $(x + 3)(x + 5)$ 26. $(m + 6)(m + 5)$

28. $(x - 2)(x - 7)$ 30. $(c + 2)(c + 20)$ 32. $(x + 6)(x - 2)$ 34. $(q + 4)(q - 7)$

36. $(5x + 2)(x + 6)$ 38. $(2x + 9)(4x - 5)$ 40. $(4m - 1)(2m - 3)$ 42. $(4g + 1)^2$

44. $(3m - 5)(2m - 1)$ 46. $3(5t - 2)(t + 5)$ 48. $(6a - 5b)(2a + 3b)$

50. $2(12m + 5)(m + 1)$ 52. $8(x + 12y)(x - 15y)$ 54. $(3x + 7)^2$ 56. $(2a^2 + 3b)^2$

58. $(3c^2 - 2y^2)^2$ 60. $(x^2 + 4)(x + 2)(x - 2)$ 62. $(b^2 + 25)(b + 2)(b - 2)$

64. $(x^2 + 3)(x + 2)(x - 2)$

Section 5.3

2. $(x + 4)(x - 4)(m - 3)$ 4. $(b + 3)(a - c)$ 6. $(x + a)(x + y)$

8. $(x + y)(x - 2)(x^2 + 2x + 4)$ 10. $(x - y + 4a^2)(x - y - 4a^2)$

12. $(3a - 12 + 8b^2)(3a - 12 - 8b^2)$ 14. $(x + 5 + 6y)(x + 5 - 6y)$

16. $(2 + a + b)(2 - a - b)$ 18. $(3x - 3y - 2)(x - y - 4)$ 20. $(x + y + z)^2$

22. $(3x + 2y)(9x^2 - 6xy + 4y^2)$ 24. $(3x - 2y)(9x^2 + 6xy + 4y^2)$

26. $(x - yz)(x^2 + xyz + y^2z^2)$ 28. $(2a + b)(4a^2 - 2ab + b^2)$

30. $(4x^2 - 5y)(16x^4 + 20x^2y + 25y^2)$ 32. $(2ab^2 + 5c^4)(4a^2b^4 - 10ab^2c^4 + 25c^8)$

34. $(ab + c)(a^2b^2 - abc + c^2)(ab - c)(a^2b^2 + abc + c^2)$

Section 5.4

2. $\frac{3x}{4}$ 4. $\frac{9z^4}{19x^3y^4}$ 6. $3x(2x - 1)^4$ 8. $\frac{4}{(2x+5)(3x-4)^2}$ 10. $\frac{x}{2}$

12. $\frac{2-t}{4-t}$ 14. $\frac{x+7}{2(x-2)}$ 16. $\frac{x-6}{2x-1}$ 18. $\frac{2x+5}{2x-3}$ 20. $-\frac{1}{2}$

22. $\frac{-x(2x+1)}{3x+4}$ 24. $\frac{2c+3}{2c-3}$ 26. $-\frac{1}{y}$ 28. $\frac{2x-1}{2x+1}$ 30. $\frac{a+3}{a}$

32. $\frac{b(a-4)}{a+5}$ 34. $\frac{x^2+xy+y^2}{x+y}$ 36. $\frac{x^2+4}{x^2+2}$ 38. $\frac{-x-y}{x-y}$

40. $\frac{-x+y-z}{x+y+z}$

Section 5.5

2. $\frac{1}{10}$ 4. 2 6. $\frac{8t^2}{9}$ 8. $\frac{3}{c^4}$ 10. $\frac{9n}{m^4p^3q}$ 12. $\frac{-5x}{a}$

14. $\frac{243x^2}{64y^9}$ 16. $\frac{2x}{3}$ 18. 32 20. $\frac{9(a+1)}{a(a-1)}$ 22. $-3s(s + 3)$ 24. $\frac{-2(y+5)}{y}$

26. $\frac{x-y}{x-1}$ 28. $\frac{x(x+y)}{z+y}$ 30. $\frac{t-5}{(t+5)^2}$ 32. $\frac{x-3}{x-2}$ 34. $\frac{5x+3}{x-4}$

36. $\frac{8m-3n}{2m-3n}$ 38. $\frac{-1}{y}$ 40. $(5ab - 7)(4ab + 3)$ 42. $x - y$ 44. $2x + y$

46. $\dfrac{x^2-4x+16}{x^2-3x+9}$ 48. $\dfrac{(2x-y)(x+1)}{(2x+1)^2}$

Section 5.6

2. $60m$ 4. $12t^3$ 6. $48a^2b^3c^4$ 8. $3x(x+3)$ 10. $12(a-3)$

12. $(b-4)(b+4)2$ 14. $(2x+1)(3x-1)(3x-4)$ 16. $90a(a+5)^2(a-5)$

18. $t^2(t+3)^2(t-3)^2$ 20. $\dfrac{3}{10}$ 22. $\dfrac{2x}{a}$ 24. $\dfrac{-2a}{2x+3}$ 26. $\dfrac{11}{40}$ 28. $\dfrac{16a+7b}{24}$

30. $\dfrac{2y+3}{3y^2}$ 32. $\dfrac{-11}{12t}$ 34. $\dfrac{20b^2-18ab^3-3a}{12a^2b^3}$ 36. $\dfrac{a^2+bc+ab}{ab}$

38. $\dfrac{27y-2x+1}{3y}$ 40. $\dfrac{3s+3}{s(s-3)}$ 42. $\dfrac{7x+15}{(x-5)(x+5)}$ 44. $\dfrac{-2a-8}{(a+3)(a+2)}$

46. $\dfrac{11t^2+4t}{(t+4)(t-2)}$ 48. $\dfrac{-c}{c-6}$ 50. $\dfrac{11}{15(r-5)}$ 52. $\dfrac{2x-5}{x-6}$

54. $\dfrac{x+2}{x+1}$ 56. $\dfrac{y+13}{(y-2)(y-3)(y+1)}$ 58. $\dfrac{5x-2}{3x(x+2)(x-2)}$ 60. $\dfrac{-2}{(x-y)(x-2y)}$

62. $\dfrac{84x}{(3x-7)(3x+7)(2x+5)}$ 64. $\dfrac{10x}{(x+2)(x-2)}$ 66. $\dfrac{-(x-1)^2}{(2x-1)(x+1)}$

68. $\dfrac{x+5}{4(x+1)^2}$ 70. $\dfrac{3}{(c+1)(c-1)(c^2+c+1)}$ 72. $\dfrac{2x}{x^2+x+1}$

Section 5.7

2. $\dfrac{8r}{r-3}$ 4. $\dfrac{x}{12}$ 6. $\dfrac{1}{4x}$ 8. $-\dfrac{13}{3}$ 10. a 12. $\dfrac{a^2-a+9}{a^2}$

14. $-\dfrac{x}{z}$ 16. $\dfrac{-1}{y+1}$ 18. $\dfrac{2-c}{2+c}$ 20. $\dfrac{x}{x-2y}$ 22. $\dfrac{2x+1}{3x-7}$

24. $\dfrac{x(4x-7)(4x+5)}{2x+5}$ 26. -1 28. 1 30. $\dfrac{x+3}{x+2}$ 32. $\dfrac{13y-6}{4y-2}$

34. $\dfrac{2x+3}{x+1}$ 36. $\dfrac{2x^2-4x+18}{x(x-6)}$

Section 5.8

2. 24   4. 9   6. 9   8. −10   10. 47   12. 7   14. 9

16. $-\dfrac{104}{11}$ 18. $\rho = \dfrac{RA}{L}$ 20. $f = \dfrac{pq}{p+q}$ 22. $R2 = \dfrac{R_1R_3 - RR_3 - RR_1}{RR_1R_3}$

24. $x = \dfrac{y}{1-y}$ 26. $x = \dfrac{a(a+1)}{1-2a}$ 28. $R2 = \dfrac{QR_1}{Q-VR_1}$ 30. $\dfrac{3}{2}$ 32. $-\dfrac{12}{7}$

34. $\dfrac{12}{7}$ 36. 2   38. −2   40. No solution   42. −1   44. No solution

CHAPTER 6

Section 6.1

2. $(-2, -3)$ 4. $(3, 0)$ 6. System inconsistent, lines parallel 8. $(5, -4)$

10. $(-2, 5)$ 12. $(-2, -3/2)$ 14. $(0, 2)$ 16. $(1, -2)$

18. System dependent, lines coincide 20. $(1, 7)$ 22. $(2, 2)$ 24. $(10/3, 5/4)$

26. $(-6, -3)$ 28. $(2, -10)$ 30. $(19, 5)$

32. $(-4, -7)$ 34. $(5, -2)$ 36. $(4, 15)$ 38. $(-15, 45)$ 40. $(18.5, 21.8)$

42. $(4.2, 6.7)$ 44. 35°, 55° 46. 20 mH, 70 mH 48. A: 40 gal B: 60 gal

50. 120 km/h, 240 km/h 52. 21 cm, 33 cm, 33 cm 54. 4@ 0.25 A; 2@ 0.45 A

56. Credit union: \$8000; bank: \$32,000 58. $\dfrac{9\sqrt{10}}{10}$

Section 6.2

2. $(a, -b)$ 4. (b^2, a^2, b^2) 6. $\left(\dfrac{1}{a+b}, \dfrac{1}{a+b}\right)$ 8. $\left(\dfrac{b^2-a^2}{a^2+b^2}, \dfrac{(b-a)^2}{a^2+b^2}\right)$

10. $(a+b, a-b)$ 12. $(3, -5)$ 14. $\left(\dfrac{1}{4}, \dfrac{1}{3}\right)$ 16. $\left(-\dfrac{3}{4}, \dfrac{1}{3}\right)$ 18. $(3/8, 9/4)$

20. $\left(\dfrac{a^2-b^2}{2a-b}, \dfrac{a^2-b^2}{a-2b}\right)$

Section 6.3

2. $(8, -5, 6)$ 4. $(-2, -7, -3)$ 6. $(4, 4, 4)$ 8. $(-6, 0, 2)$ 10. $(5, 2, 2)$

12. $(3, -4, \dfrac{3}{4})$ 14. $(-6, 3, 4, -2)$ 16. (2 A, 3 A, 5 A)

18. 15 acres at \$2700, 20 acres at \$2200, 40 acres at \$1450

20. 16 L of 20%, 32 L of 30%, 32 L of 60%

Section 6.4

2. 34 4. –23 6. –1 8. –28 10. –35 12. 18 14. –44 16. –30

18. $m^2n^2 + mn$ 20. –16 22. 173 24. 102 26. 0 28. 0 30. 84

32. –3 34. 1 36. –512

Section 6.5

2. 0 4. 0 6. –145 8. 54 10. 210 12. 30 14. –241 16. –8

18. 123 20. 1170 22. –596 24. 32,562

Section 6.6

2. (7, 3) 4. (–1, –5) 6. (3, 0) 8. Dependent; lines coincide

10. Inconsistent; lines parallel 12. (–11/4, –9/4) 14. (–3/5, 1/5)

16. $\frac{3a-2b}{3b-2a}, \frac{a^2-b^2}{3b-2a}$ 18. (0, –1) 20. $\frac{a^2}{a^2-b}, \frac{a}{a^2-b}$

22. 35 kg of 25% mixture, 21 kg of 65% mixture 24. (1, –5, 6) 26. (5, 7, 2)

28. (–2, –5, 7) 30. (5/4, –2/3, –9/8) 32. (3, 7, –3) 34. (0, –4, –7/2)

36. $I_1 = 4.25$ A, $I_2 = 6.50$ A, $I_3 = 2.25$ A 38. (4, –2, 3, 1) 40. (1, 2, 0, 2, –1)

Section 6.7

2. $\frac{3}{x-4} + \frac{7}{x-2}$ 4. $\frac{7}{x+1} - \frac{4}{3x-2}$ 6. $\frac{6}{x+1} - \frac{3}{2x+3} + \frac{1}{x-3}$

8. $\frac{3}{2x-1} - \frac{6}{(x-1)^2}$ 10. $\frac{1}{x+4} - \frac{16}{(x+4)^3}$ 12. $\frac{1}{x} + \frac{2}{x^2} + \frac{5}{x+2}$

14. $\frac{2}{2x^2+1} - \frac{3x}{x^2+5}$ 16. $\frac{x}{3x^2-x-1} + \frac{2}{x^2+4}$ 18. $\frac{4}{x^2+x-1} + \frac{1}{x-4} - \frac{1}{x+4}$

20. $\frac{2}{x-1} - \frac{6x}{x^2-3} + \frac{2x+1}{(x^2-3)^2}$ 22. $\frac{3}{x} + \frac{1}{x^2} + \frac{2}{(x^2+4)^2}$ 24. $\frac{1}{x^2} - \frac{1}{(x^2+1)^2} + \frac{1}{(x^2+1)^3}$

26. $x^2 + x + 4 - \frac{2/3}{x+1} + \frac{20/3}{x+2}$ 28. $2x + \frac{3}{x} + \frac{5}{x-1}$ 30. $x^2 - 3x + 1 + \frac{3}{x+2} + \frac{4x+5}{x^2+2}$

CHAPTER 7

Section 7.1

2. $-5, -\frac{1}{3}$ 4. $3, 6$ 6. $-6, 7$ 8. $-5, \frac{1}{2}$ 10. $\frac{4}{9}, \frac{1}{4}$ 12. $\frac{8}{7}, -\frac{3}{2}$

14. $0, \frac{2}{3}$ 16. $1, \frac{1}{2}$ 18. $4, -4$ 20. $\pm\frac{3}{2}$ 22. $\pm\frac{2\sqrt{6}}{3}$ 24. $\pm\frac{2\sqrt{7}}{7}$

26. $\pm 9\sqrt{2}$ 28. $\pm\frac{\sqrt{3}}{6}$ 30. $\frac{4}{3}, \frac{3}{4}$ 32. $\frac{1}{3}, -2$ 34. $\frac{4}{3}, 4$ 36. $\pm\frac{4}{3}$

38. $\frac{11}{7}, 2$

Section 7.2

2. $-6, -9$ 4. $-3, 9$ 6. $\frac{1}{2}, 3$ 8. $-4, -\frac{5}{3}$ 10. $\frac{5 \pm \sqrt{13}}{2}$

12. $\frac{1}{5}, -1$ 14. $\frac{-5 \pm \sqrt{73}}{6}$ 16. $\frac{4 \pm \sqrt{6}}{5}$ 18. $\frac{1}{3}, -2$ 20. $\frac{7 \pm \sqrt{41}}{16}$

Section 7.3

2. $-3, -7$ 4. $\frac{1}{2}, -5$ 6. $\frac{5}{2}, -\frac{5}{3}$ 8. $4, -\frac{5}{2}$ 10. $\frac{15}{2}, -\frac{3}{2}$ 12. $\pm\sqrt{3}$

14. $\frac{1 \pm 2\sqrt{3}}{2}$ 16. $3 \pm 2\sqrt{7}$ 18. $\frac{5 \pm 2\sqrt{7}}{2}$ 20. $\frac{5 \pm 3\sqrt{29}}{6}$ 22. $\frac{-7 \pm \sqrt{41}}{4}$

24. $\frac{-2 \pm \sqrt{22}}{6}$ 26. $1, 7$ 28. $\frac{-5 \pm \sqrt{57}}{4}$ 30. $\frac{5}{3}$ 32. No real solution

34. 0.628, –2.95 36. 0.179, –1.60 38. 1.14, –0.0879 40. 1.33, 0.349

42. $3, -4; 5; \frac{-1 \pm \sqrt{73}}{2}$ 44. $0, -\frac{2}{3}; \frac{10}{3}, -4; \frac{-1 \pm \sqrt{10}}{3}$; no real solution

46. $\frac{-3n \pm n\sqrt{13}}{4m}$ 48. $\frac{-2 \pm \sqrt{4 + 3mn}}{3}$ 50. $-3n$ 52. $\frac{-\pi h \pm \sqrt{x^2h^2 + 2\pi S}}{2\pi}$

Section 7.4

2. 4, 6   4. 60.0 m, 90.0 m   6. 6.00 ft, 8.00 ft

8. 5.00 s, 9.00 s; 1.00 s, 13.0 s; 6.00 s, 8.00 s

10. 0.500 ms, 1.00 ms; $\dfrac{3 \pm \sqrt{5}}{4}$ ms or 0.191 ms, 1.31 ms; no real solutions

12. 3.00 cm, 8.00 cm   14. 12.0 m, 24.0 m, 60.0 m   16. 2.00 cm, 832 cm^2

18. (a) 1.50 s (b) 2.40 s (c) 0 s, 6.00 s (d) 144 ft

20. 5.00 in or 7.00 in.   22. 2.00 in.

24. $26\bar{0}$ ft × $43\bar{0}$ ft   26. 1.00 A, 5.00 A   28. 73.2 m

CHAPTER 8

Section 8.1

2. c^{10} 4. $\frac{1}{y^3}$ 6, z^9 8. $\frac{1}{c^2}$ 10. p^7 12. 4^9 14. $\frac{1}{r^2}$

16. $\frac{1}{7^3}$ 18. $\frac{1}{2^8}$ 20. $\frac{125}{8}$ 22. 4 24. $\frac{4}{5}$ 26. 5^2 or 25 28. $\frac{125}{27}$

30. $\frac{1}{a^5}$ 32. $\frac{1}{a^6}$ 34. a^6 36. $16a^4b^2$ 38. $\frac{y^3}{64x^6}$ 40. $\frac{24}{b^3}$ 42. $\frac{2s^3}{t^6}$

44. p^6 46. $\frac{1}{t^9}$ 48. $\frac{b^4}{a^3}$ 50. s^{12} 52. $\frac{1}{s}$ 54. $\frac{b^5}{a^2}$ 56. $\frac{1}{x^4}$

58. $\frac{a^2}{64}$ 60. $\frac{81a^6}{b^{20}c^8}$ 62. $\frac{b^9}{27a^{12}}$ 64. $\frac{1}{a+b}$ 66. $\frac{a^2+b^3}{ab^2}$ 68. $\frac{1}{4}$

70. $\frac{xy}{x+y}$ 72. 1

Section 8.2

2. 7 4. 5 6. 4 8. 4 10. $\frac{1}{5}$ 12. $\frac{1}{2}$ 14. 256 16. $\frac{1}{8}$

18. $\frac{2}{3}$ 20. $\frac{8}{27}$ 22. $\frac{9}{4}$ 24. $\frac{1}{9}$ 26. 125 28. 2 30. 2 32. $\frac{1}{m}$

34. $\frac{1}{a^{7/6}}$ 36. $\frac{1}{x^{1/4}}$ 38. x 40. $x^{1/6}$ 42. $x^{1/8}y^{9/8}$ 44. $\frac{a^{10}b^{15}}{c^{25}}$

46. $\frac{ab^2}{c^4}$ 48. $a^{1/3} - a^{3/4}$ 50. $2s^{5/4} - 6$ 52. $\frac{15}{p} - 30$ 54. $x - \frac{1}{y}$

56. $6x + x^{1/2}y^{1/2} - 12y$ 58. 48 60. 3×10^{-6} 62. 1800 64. $\frac{20}{9}$

66. 1.35×10^{-5}

Section 8.3

2. 9 4. $6\sqrt{2}$ 6. $4\sqrt{6}$ 8. $3x\sqrt{7}$ 10. $3x\sqrt{5x}$ 12. $2ab^2\sqrt{15a}$

14. $6a^2b^6c^3\sqrt{3ac}$ 16. $\frac{\sqrt{5}}{4}$ 18. $\frac{\sqrt{21}}{6}$ 20. $\frac{\sqrt{6}}{3}$ 22. $\frac{4\sqrt{2}}{5}$

24. $\frac{\sqrt{6ab}}{5b}$ 26. $\frac{\sqrt{5y}}{y}$ 28. $\frac{y^2\sqrt{3xy}}{2x^2}$ 30. 7 32. $3c\sqrt[3]{5c^2}$

34. $5c^3\sqrt[3]{2c}$ 36. $4x^2\sqrt[3]{x}$ 38. $2xy^2\sqrt[3]{9xz^2}$ 40. $\frac{\sqrt[3]{5}}{3}$ 42. $\frac{\sqrt[3]{15}}{3}$

44. $\frac{\sqrt[3]{63}}{6}$ 46. $\frac{2\sqrt[3]{12}}{3}$ 48. $\frac{\sqrt[3]{75}}{5}$ 50. $\frac{y\sqrt[3]{4x^2yz^2}}{2xz}$ 52. $\frac{\sqrt[3]{9x^2y}}{3}$

54. $3a^2\sqrt[4]{2}$ 56. $2b\sqrt[5]{2a^2b^2}$ 58. $\frac{c\sqrt[4]{12abc}}{2a}$ 60. $b\sqrt{a}$ 62. $2ab^3$

64. $\sqrt[6]{5}$ 66. $2\sqrt[4]{2}$ 68. $c\sqrt[3]{3a^2bc}$ 70. $3\sqrt[3]{a^6+b^9}$ 72. $3(a^2+b^3)$

Section 8.4

2. $12\sqrt{5}$ 4. $-2\sqrt{3}$ 6. $12\sqrt{2}$ 8. $14\sqrt{7}$ 10. $15\sqrt{3}$ 12. $-4\sqrt{2}$

14. $-19\sqrt{5}$ 16. $8\sqrt{10}$ 18. $15\sqrt{2c}$ 20. $-28t\sqrt{3}$ 22. $8\sqrt[3]{5}$ 24. $12\sqrt[3]{3}$

26. $3\sqrt[3]{4}$ 28. $5\sqrt[3]{2}-\sqrt{2}$ 30. $\sqrt[3]{6}$ 32. $\frac{2\sqrt{5}}{5}$ 34. $\frac{11\sqrt{2}}{5}$ 36. $\frac{41\sqrt{10}}{20}$

38. $\frac{2\sqrt{6}}{3}$ 40. $\frac{\sqrt{3}}{3}$ 42. $\frac{4\sqrt{21}}{3}$ 44. $\frac{\sqrt[3]{2}}{2}$ 46. $\frac{-5\sqrt[3]{12}}{6}$ 48. $\frac{23\sqrt{6}}{6}$

50. $(-x-x^2)\sqrt{3x}$ 52. $(a-b+1)\sqrt[3]{ab^2}$ 54. $(a+1-8a^2)\frac{\sqrt{2a}}{2a}$

56. $(ab-3a+3b-6a^2b^3)\frac{\sqrt{3a^2b}}{3ab}$

Section 8.5

2. $6\sqrt{5}$ 4. $6\sqrt[3]{4}$ 6. 2 8. $\sqrt[4]{8}$ 10. $2\sqrt[6]{2}$ 12. 6 14. $\sqrt[6]{32}$

16. $\frac{\sqrt[4]{72}}{6}$ 18. $33-11\sqrt{3}$ 20. 16 22. $26+4\sqrt{3}$ 24. $6a^2+11a\sqrt{ab}+3ab$

26. $28-17\sqrt{10}$ 28. $3a-7\sqrt{ab}+4b$ 30. $36+16\sqrt{5}$ 32. $73-12\sqrt{35}$

34. $9a^2 + 12a\sqrt{b} + 4b$ 36. $16a - 16\sqrt{ab} + 4b$ 38. $-2 - \sqrt{3}$ 40. $\dfrac{-16 + 9\sqrt{6}}{46}$

42. $\dfrac{14 + \sqrt{6}}{19}$ 44. $\sqrt{10}$ 46. $\dfrac{6 - \sqrt{11}}{5}$ 48. $\dfrac{a + 2\sqrt{ab} + b}{a - b}$

Section 8.6

2. $\dfrac{129}{4}$ or $32\dfrac{1}{4}$ 4. 4 6. 63 8. 33 10. 8 12. 2, $\dfrac{14}{3}$ 14. $-\dfrac{3}{4}$, 2

16. 3, $\dfrac{7}{15}$ 18. −2 20. −1 22. −3 24. 0, 7 26. 25 28. $r = \dfrac{2GM}{v^2}$

30. $m = \dfrac{RT}{v^2 N}$ 32. $g = \dfrac{w\sqrt{w^2{}_n - w_d{}^2}}{c}$ 34. $X_L = X_C + \dfrac{\sqrt{E^2 - I^2R^2}}{I}$ 36. $R_2 = \dfrac{R_1T^2}{4\pi^2C - T^2}$

38. $r = \dfrac{GmD}{Gm - v^2D}$ 40. 5, 4

Section 8.7

2. 2, −2, $\sqrt{6}, -\sqrt{6}$ 4. $\dfrac{1}{2}, -\dfrac{1}{2}, \dfrac{1}{3}, -\dfrac{1}{3}$ 6. $-\dfrac{4}{3}, -\dfrac{6}{5}$ 8. 2, 3, $\dfrac{5}{2}$ 10. $\dfrac{1}{4}$

12. $\dfrac{1 \pm \sqrt{3}}{2}$ 14. 64, −27 16. 9, 28 18. $\pm\sqrt{\dfrac{3 \pm \sqrt{3}}{3}}$ or ±1.26; ±0.650 20. 1

CHAPTER 9

Section 9.1

2.

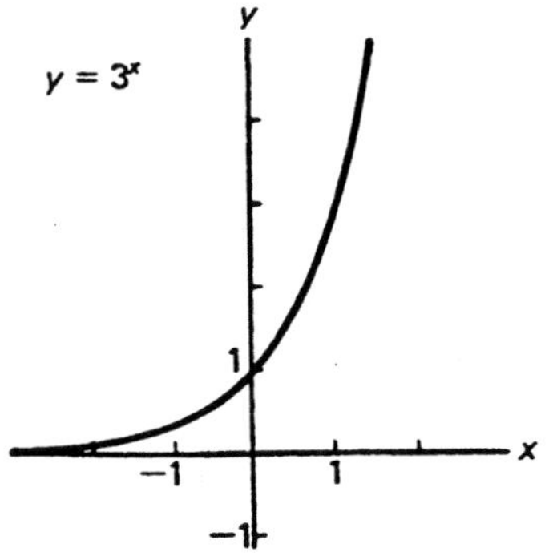

4.

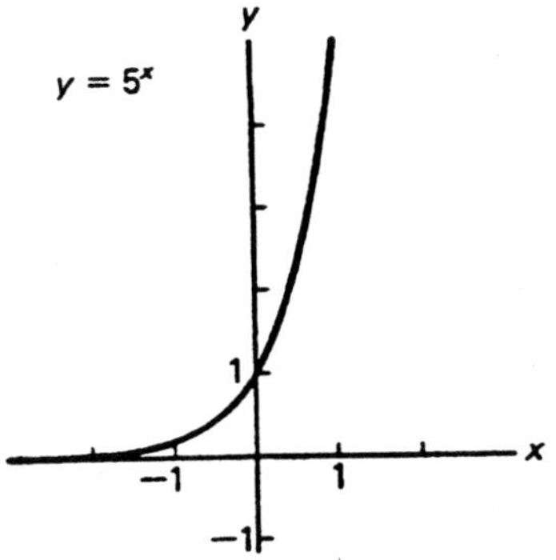

6.

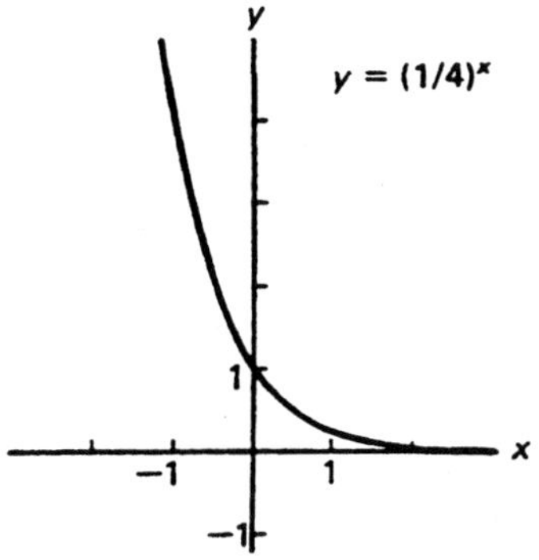

8.

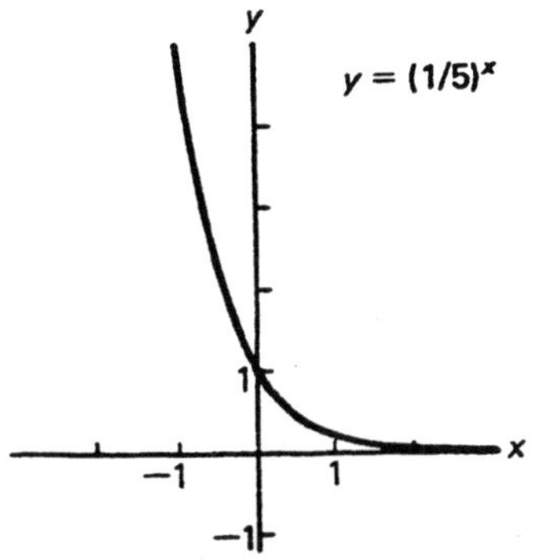

10.

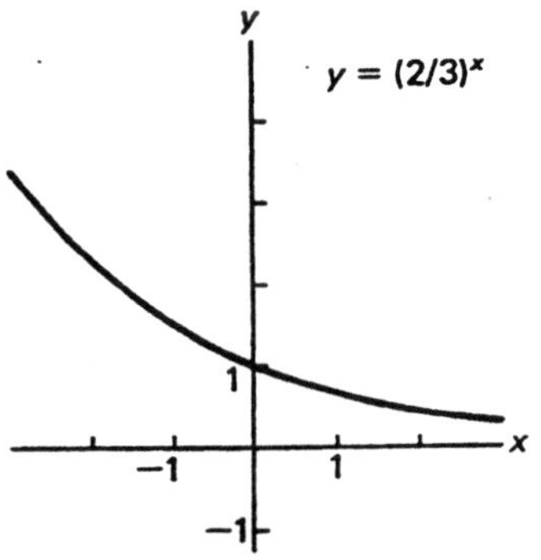

12.

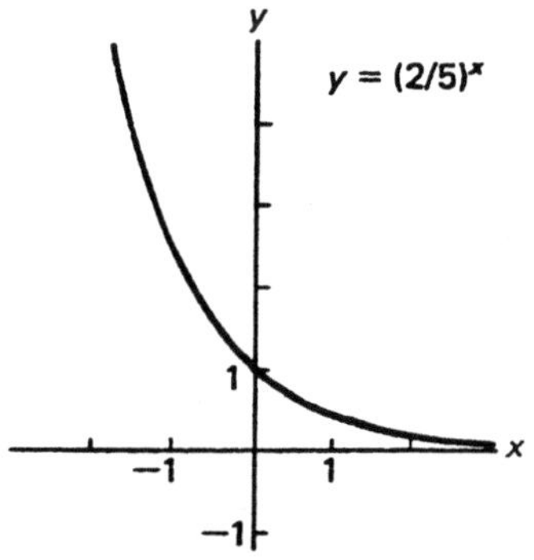

14.

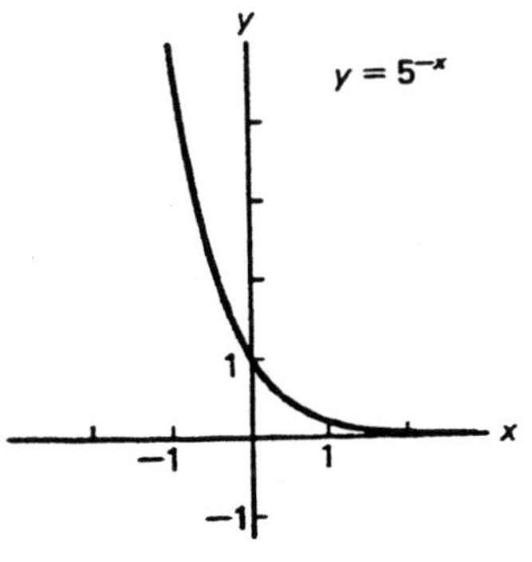

16.

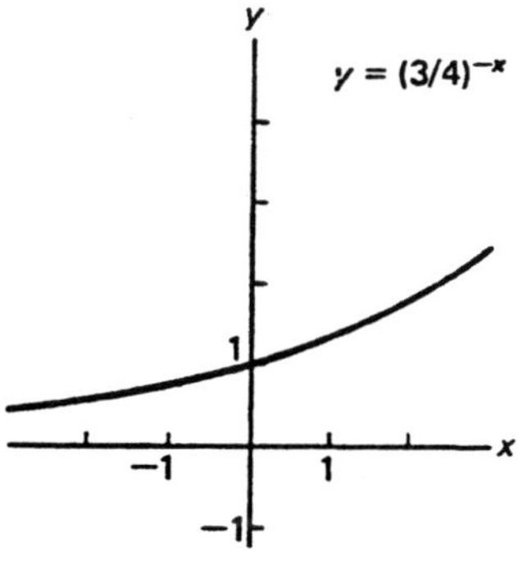

18.

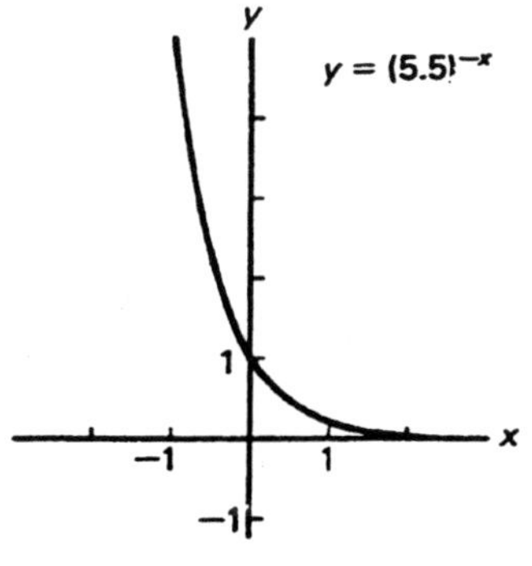

20.

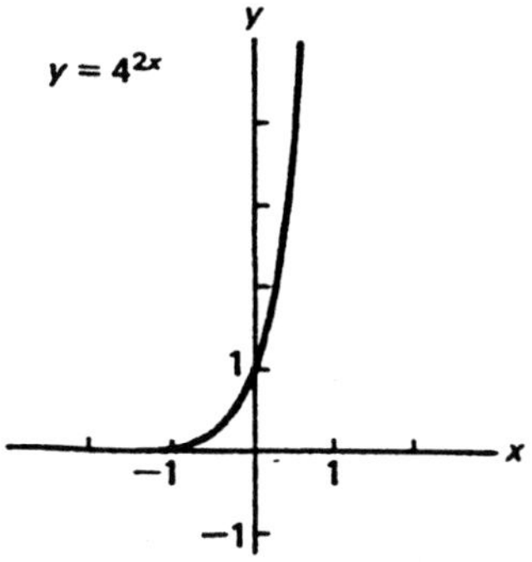

22.

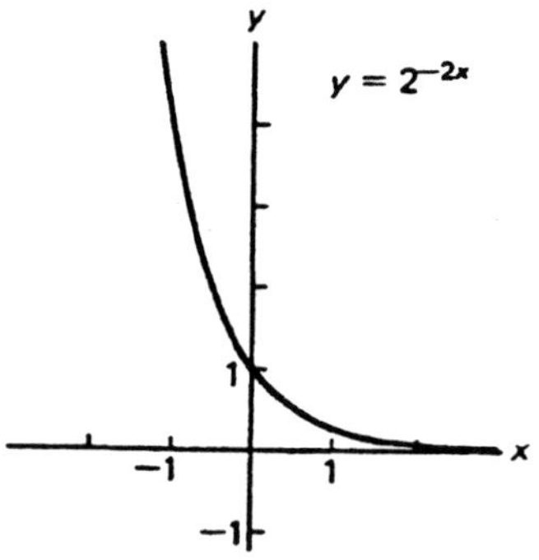

24.

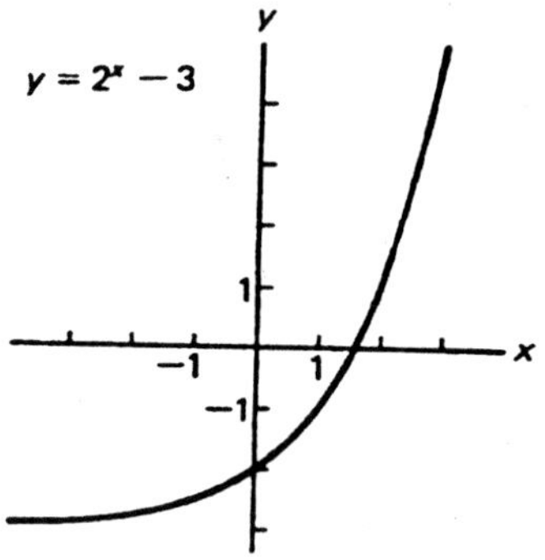

26. 2.11 28. 19.4 30. 77.9 32. 3.30 34. 8.39 36. 2.29

38. 1.55 40. 36.0 42. \$33,000 44. (a) \$1093.44 (b) \$470.49 46. 0.636

Section 9.2

2. $\log_7 49 + 2$ 4. $\log_{10} 1000 = 3$ 6. $\log_3 81 = 4$ 8. $\log_{16} 4 = \frac{1}{2}$

10. $\log_4 1 = 0$ 12. $\log_d f = e$ 14. $8^2 = 64$ 16. $5^3 = 125$ 18. $81^{1/2} = 9$

20. $27^{1/3} = 3$ 22. $2^{-3} = \frac{1}{8}$ 24. $g^k = h$

26.

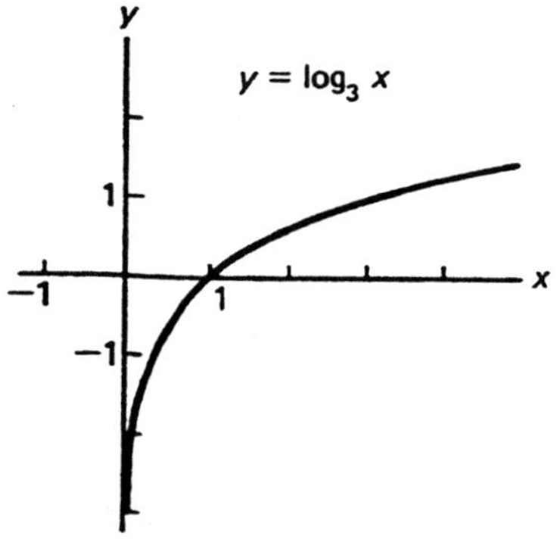

28.

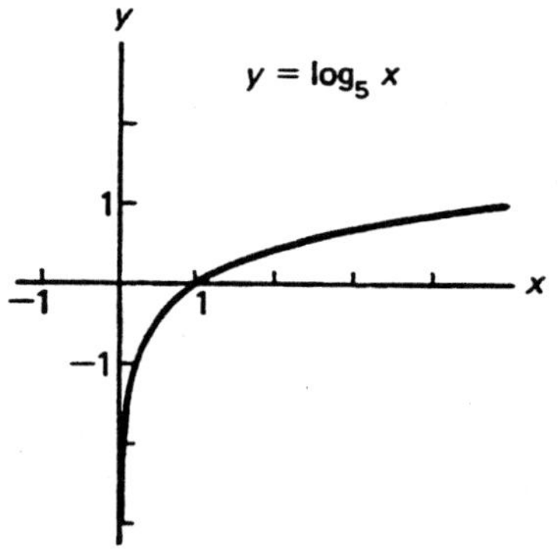

30.

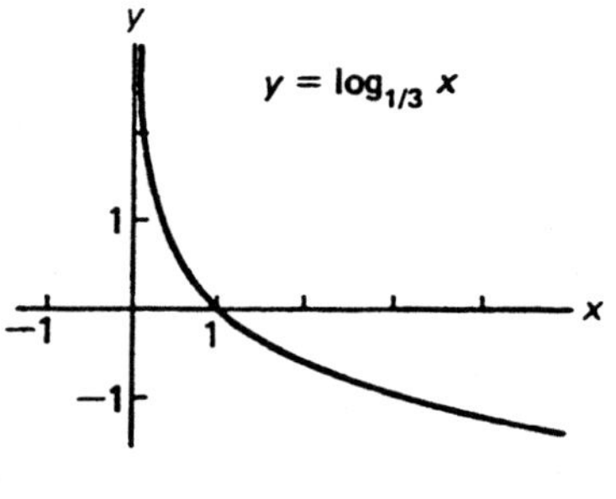

32. $\frac{1}{2}$ 34. 2 36. 3 38. $\frac{1}{3}$ 40. 3 42. 9 44. −2 46. 16

48. 81

Section 9.3

2. $\log_3 8 + 2\log_3 x + 3\log_3 y - 4\log_3 z$
4. $3\log_4 y - \log_4 x - \frac{1}{2}\log_4 z$
6. $\log_b 7 + \log_b x + \log_b y - \frac{1}{3}\log_b z$
8. $\frac{1}{4}\log_5 x + \frac{1}{2}\log_5 y + \frac{1}{4}\log_5 z$
10. $\frac{2}{3}\log_b x - \frac{1}{3}\log_b y - 2\log_b z$
12. $\frac{1}{2}\log_b y + \frac{1}{4}\log_b x - 2\log_b z$
14. $-\frac{3}{2}\log_b x + \frac{1}{2}\log_b (x+4)$
16. $\log_b \frac{z^2}{x^3}$
18. $\log_7 \frac{x^3}{y^4 z^5}$
20. $\log_2 \frac{\sqrt{x}}{z\sqrt[3]{y}}$
22. $\log_3 \frac{(x+1)\sqrt{x+2}}{(x-1)^3}$
24. $\log_b \frac{(x+1)^3\sqrt{x-7}}{x^2}$
26. $\log_b \frac{\sqrt{x+1}}{[x(x-1)(2x-1)]^3}$
28. 5
30. 4
32. 3
34. –3
36. –1
38. 0
40. 9
42. 8
44. $\frac{1}{27}$
46. 6

Section 9.4

2. 2.968
4. 4.146
6. –1.585
8. –4.304
10. –0.2161
12. 0
14. 416
16. $55,\bar{0}00$
18. 3.71
20. 0.00183
22. 1.57×10^{-5}
24. 6.55×10^{-6}
26. 4.97
28. 0.781
30. 0.823
32. 1.97
34. 0.786
36. 2.95
38. 1.29
40. –5.15
42. 8
44. 5.7
46. 2.2
48. 20 db
50. 70 db
52. 17 db
54. –0.79 db

Section 9.5

2. 20.1
4. $22,\bar{0}00$
6. 0.0498
8. 0.000454
10. 8.17
12. 2.12
14. 0.247
16. 0.522
18. 1.87
20. 0.435
22. 4.522
24. 7.520
26. –0.3496
28. –13.68
30. 4.98
32. 0.864
34. 0.0231
36. 1.39×10^{-11}
38. $5400
40. 8.60%
42. 11,400
44. 150,000
46. 0 g(1.15×10^{-50} g)
48. 0.0159 A
50. 160 yr
52. 1700 yr
54. 45,000 Btu/h
56. 25.5 V
58. –63.1 V
60. 0.065 cm/min
62. 8.21

64. 0.702 66. 1.03 68. 1.44

Section 9.6

2. 2.26 4. −0.594 6. −0.954 8. 3.11 10. −9.744 12. 0.1302

14. 0.1303 16. −2.688 18. 0.08399 20. 2.704 22. 11.56 24. −0.5596

26. −0.5352 28. 1.299 30. 0.5512 32. ±1.915 34. −0.5108

36. 1.253 38. −2.177 40. 6.931 42. −1.073 44. 1.100 46. −0.02242

48. 0.2639 50. 0.2058 52. 10.4μs 54. 42.8 ms 56. −55.9 V

Section 9.7

2. 3 4. 25 6. 9 8. 2 10. 5 12. 13 14. 2 16. $\frac{3}{2}$

18. 1.649 20. 48 22. 8.389 24. 41.5 26. 1.180 28. 13.28

30. 18.02 32. 1.317 34. 3.938 36. $\frac{3e}{2}$ or 4.077

Section 9.8

2. (a) 16.7 yr (b) 12.6 yr 4. 11 h 6. 2.5×10^{-5} M/L 8. 7.9×10^{-9} M/L

10. 3.2×10^{-7} M/L 12. 1.0×10^{-10} W/cm^2 14. 1.0×10^{-5} W/cm^2

16. 1.0×10^{-6} W/cm^2 18. (a) 3.3 yr (b) 8.7 yr 20. 0.012 s 22. 0.69 ms

24. 14.7 H 26. 2.1 mg 28. 49.5 mg 30. 40.7 mg

Section 9.9

2. $y = 8.3x - 40$ 4. $y = 0.42x + 0.18$

6.

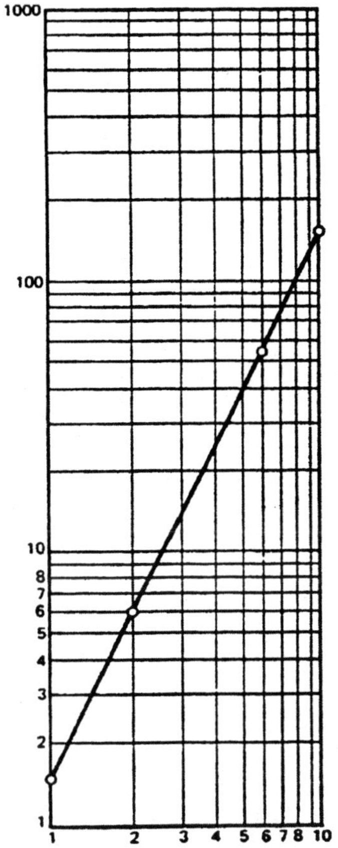

8.

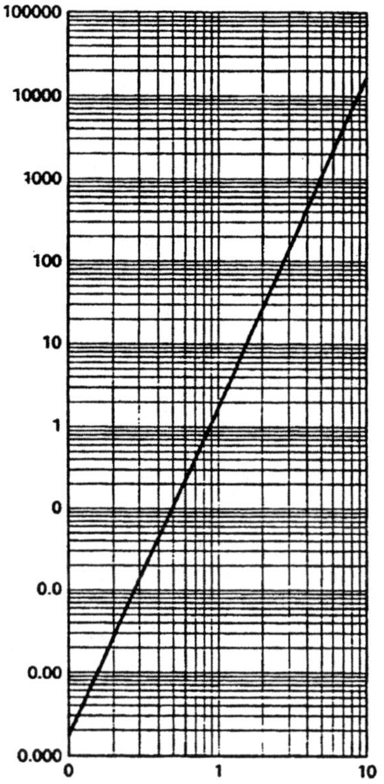

10. $y = 3.5x^2$

12. $y = 2.3x^2$

20. $y = 3.6^x$

22. $y = 1.2^x$

24. $y = 1.6(2.7)^x$

14.

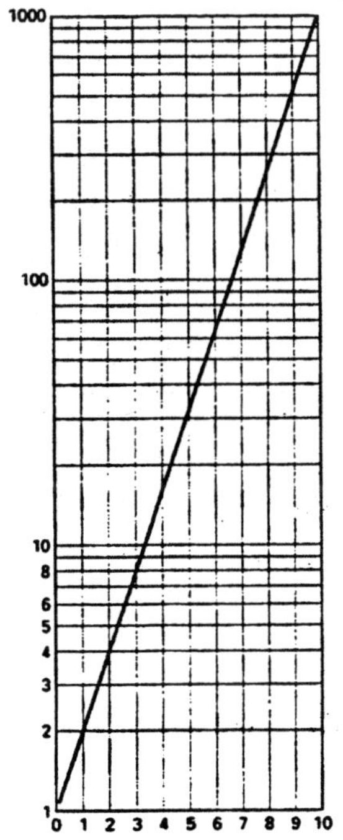

16.

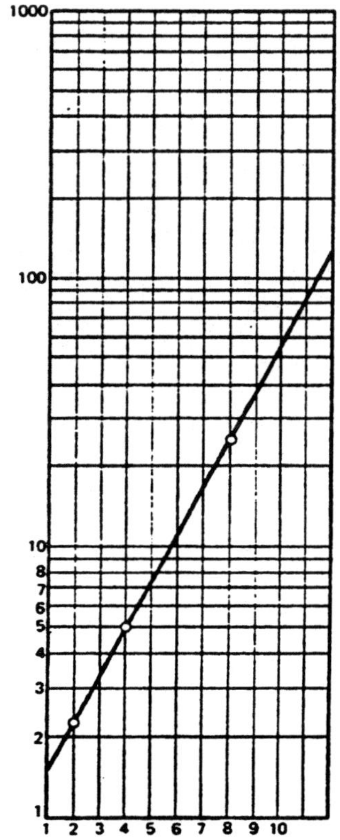

18.

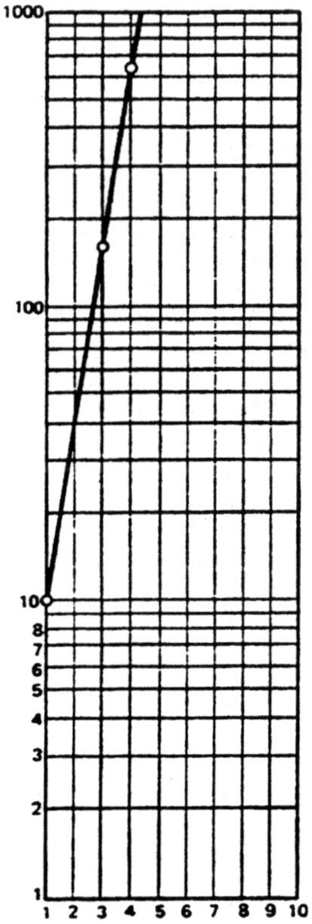

CHAPTER 10

Section 10.1

2.

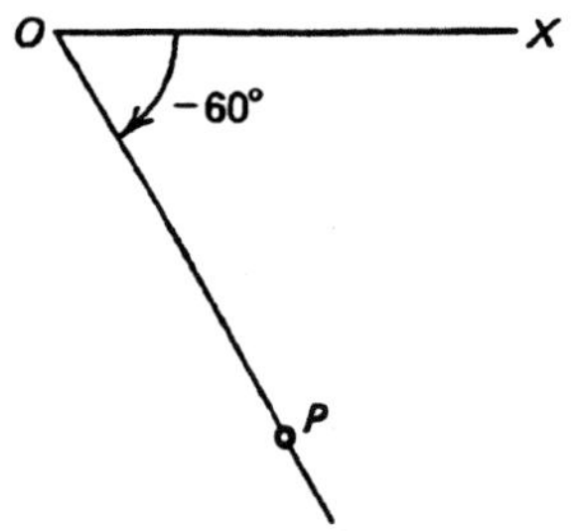

4.

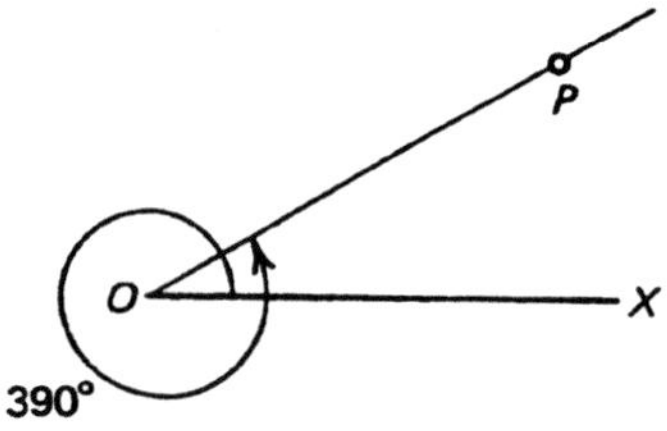

6.

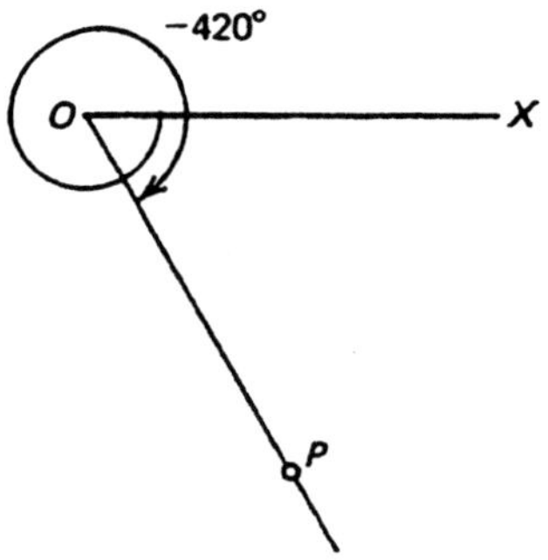

8. 535°, −185° 10. 90°, −630° 12. 660°, −60° 14. 220°, −140°

16.

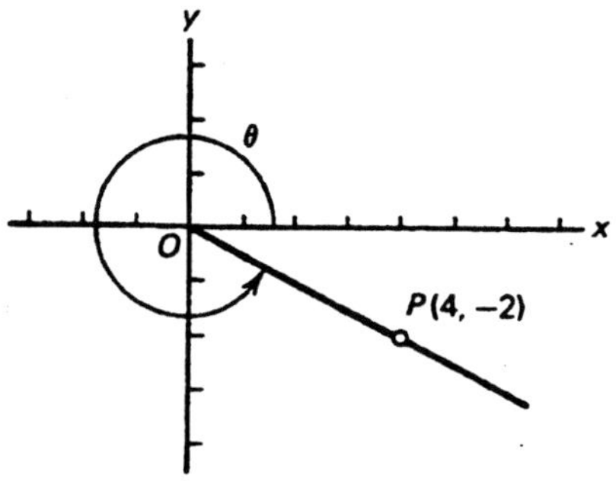

18.

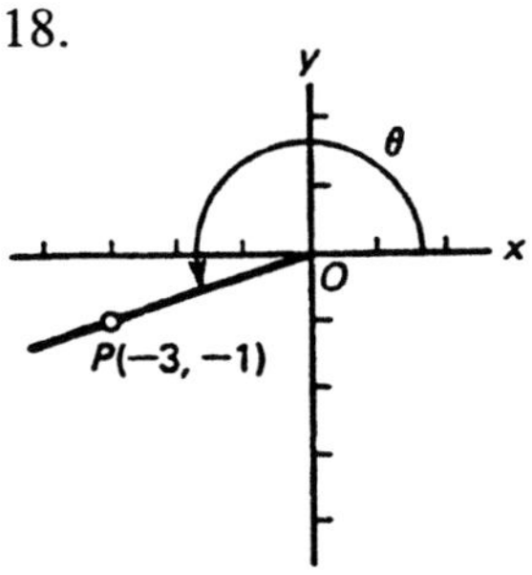

20.

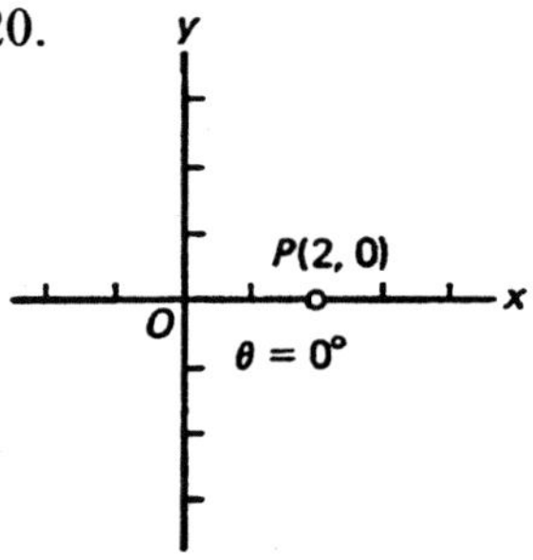

	$\sin\theta$	$\cos\theta$	$\tan\theta$	$\cot\theta$	$\sec\theta$	$\csc\theta$
22.	−3/5	−4/5	3/4	4/3	−5/4	−5/3
24.	$1/\sqrt{2}$	$-1/\sqrt{2}$	−1	−1	$-\sqrt{2}$	$\sqrt{2}$
26.	12/13	5/13	12/5	5/12	13/5	13/12
28.	$-1/\sqrt{10}$	$3/\sqrt{10}$	−1/3	−3	$\sqrt{10}/3$	$-\sqrt{10}$
30.	0	1	0	undefined	1	undefined

32. $\sin\theta = \dfrac{6}{3\sqrt{5}} = \dfrac{18}{9\sqrt{5}}$ $\cos\theta = \dfrac{3}{3\sqrt{5}} = \dfrac{9}{9\sqrt{5}}$ $\tan\theta = \dfrac{6}{3} = \dfrac{18}{9}$

$\cot\theta = \dfrac{3}{6} = \dfrac{9}{18}$ $\sec\theta = \dfrac{3\sqrt{5}}{3} = \dfrac{9\sqrt{5}}{9}$ $\csc\theta = \dfrac{3\sqrt{5}}{6} = \dfrac{9\sqrt{5}}{18}$

34. $\dfrac{1}{\cot\theta} = \dfrac{1}{x/y} = \dfrac{y}{x} = \tan\theta$ 36. $\dfrac{1}{\sin\theta} = \dfrac{1}{y/r} = \dfrac{r}{y} = \csc\theta$

38. $\dfrac{\cos\theta}{\sin\theta} = \dfrac{x/r}{y/r} = \dfrac{x}{y} = \cot\theta$

Section 10.2

2. 48° 4. 19° 6. 12.5° 8. 85.2° 10. 57°13′ 12. 26°

14. –0.7944 16. –6.855 18. –0.9143 20. 0.9952 22. –1.503

24. –3.864 26. 79.3°, 280.7° 28. 118.3°, 298.3° 30. 241.2°, 298.8°

32. 30.1°, 149.9° 34. 204.5°, 335.5° 36. 69.2°, 290.8° 38. 97.8°, 277.8°

40. 211.5°, 328.5° 42. 67.1°, 247.1°

Section 10.3

2. $\frac{7\pi}{6}$ 4. $\frac{3\pi}{2}$ 6. $\frac{-23\pi}{18}$ 8. $\frac{83\pi}{6}$ 10. 210° 12. 607.5°

14. 2160° 16. 104.3° 18. $\frac{\pi}{6}$ 20. $\frac{\pi}{3}$ 22. $\frac{5\pi}{12}$ 24. $\frac{\pi}{7}$ 26. $\frac{-\sqrt{3}}{2}$

28. $\sqrt{2}$ 30. –1 32. $\frac{\pi}{3}, \frac{2\pi}{3}$ 34. $\frac{3\pi}{4}, \frac{7\pi}{4}$ 36. $\frac{4\pi}{3}, \frac{5\pi}{3}$ 38. 0.9211

40. 0.1511 42. 0.6816 44. –0.5135 46. –0.03807 48. –10.99

50. –0.3827 52. 0 54. –0.5000 56. 1.207, 5.076 58. 3.893, 5.532

60. 2.978, 6.119 62. 0.1887, 2.953 64. 1.823, 4.460 66. 1.677, 4.819

68. 2.618, 5.760 70. 2.356, 5.498 72. 3.665, 5.760

Section 10.4

2. 6.70 in. 4. 26.8 in^2 6. 10.2 in. 8. 2.00 rad or 115.0° 10. 115 mi

12. 206 hp 14. (a) $23\bar{0}$ rad/s (b) $69\bar{0}$ rad of $11\bar{0}$ rev (c) 1380 ft/s

16. (a) 7.0 rad/s^2 (b) 8.4 rad/s^2 18. 29.9 km/s 20. (a) 74.5 rad/s (b) 119 rev

(c) 807 ft

CHAPTER 11

Section 11.1

2. $A = 72.4°$, $B = 50.1°$, $a = 188$ mi   4. $B = 146.0°$, $C = 14.5°$, $b = 816$ km

6. $A = 76.5°$, $a = 6680$ m, $c = 5920$ m   8. $B = 14.9°$, $C = 148.7°$, $c = 377$ ft

10. $B = 85.27°$, $C = 35.78°$, $b = 4507$ m   12. $C = 96.27°$, $b = 10.92$ km, $c = 14.46$ km

14. $A = 102°$, $C = 36°$, $a = 7.7$ km   16. $B = 22°$, $C = 53°$, $b = 180$ mi

18. $B = 29°33'2''$, $C = 100°\ 55'8''$, $c = 48, 480$ ft   20. $A = 34°17'36''$, $a = 55.96$ m, $b = 96.16$ m

22. (a) $32°10'12''$ (b) 3868 m   24. (a) 21.9 ft (b) 20.0°

Section 11.2

2. $A = 59.0°$, $C = 66.0°$, $a = 19\overline{0}$ m or $A = 11.0°$, $C = 114.0°$, $a = 42.4$ m

4. No triangle   6. $A = 147.2°$, $B = 15.6°$, $a = 4.03$ m   8. $B = 54.9°$, $C = 26.3°$, $c = 317$ ft

10. $A = 48°$, $C = 96°$, $c = 130$ cm or $A = 132°$, $C = 12°$, $c = 27$ cm

12. $B = 128°$, $C = 22°$, $b = 7600$ ft   14. $A = 101°$, $C = 34°$, $a = 4.4$ m

16. $B = 149.75°$, $C = 14.50°$, $b = 1192$ m

18. $A = 72.03°$, $C = 57.97°$, $a = 51{,}200$ km or $A = 7.97°$, $C = 122.03°$, $a = 7466$ km

20. $B = 34.40°$, $C = 114.10°$, $c = 655.1°$ mm or $B = 145.60°$, $C = 2.90°$, $c = 36.31$ mm

22. $B = 78°48'9''$, $C = 39°59'21''$, $b = 387\overline{0}$ ft

24. $B = 24°51'31''$, $C = 133°23'29''$, $c = 350\overline{0}$ m or $B = 155°8'29''$, $C = 3°6'31''$, $c = 261.2$ m

Section 11.3

2. $A = 137.3°$, $C = 23.2°$, $b = 10.6$ ft   4. $B = 11.7°$, $C = 73.6°$, $a = 909$ yd

6. $A = 39.8°$, $B = 92.2°$, $C = 48.0°$ 8. $A = 13.4°$, $B = 149.2°$, $C = 17.4°$

10. $A = 44°$, $C = 76°$, $b = 2\overline{0}0$ cm   12. $A = B = 34°$, $C = 112°$

14. $B = 15°$, $C = 16\overline{0}$, $a = 6.3$ m   16. $A = 45.59°$, $B = 59.41°$, $c = 50{,}770$ mi

18. $B = 24.97°$, $c = 43.98°$, $a = 4368$ ft   20. $A = 30.55°$, $B = 107.45°$, $C = 42.00°$

22. $A = 49°53'53''$, $B = 32°40'37''$, $c = 935.0$ cm   24. $A = 78°43'17''$, $B = 48°23'41''$, $C = 52°53'2''$

26. (a) 201.6 m (b) 29°37′27″ 28. (a) 13.9 ft (b) 68.0°

Section 11.4

2. 89.5° 4. 23.7 ft 6. 6900 ft 8. 305 m 10. 146 ft, 207 ft

12. (a) 846 ft (b) 227 ft 14. $L1 = 19.6$ ft, $L2 = 11.7$ ft 16. 2.3×10^8 km or 4.6×10^7 km

18. (a) 75.2° (b) 60.8° (c) 5.42 m (d) 7.80 m 20. 173 m

22. (a) 362 ft (b) $28\bar{0}$ ft (c) 344 ft 24. 70.8° 26. 20.3 mi

Section 11.5

2. $4\bar{0}0$ mi at 201° 4. 68 km/h at 39° 6. 660 km at 359° 8. 110 km at 295°

10. 180 mi at 247° 12. 340 mi/h at 222° (48° west of south)

Section 11.6

2. $64\bar{0}0$ m at 286.1° 4. 397 mi at 200.9° 6. 115 km/h at 270.0°

8. 42.1 mi at 125.0° 10. 21.4 mi at 36.8° 12. 109 mi at 43.4° (43.4° north of east)

14. 304 km at 231.7° (38.3° west of south) 16. 226 mi/h at 101.8° (11.8° west of north)

Section 11.7

2. $\mathbf{v}x = -20.7$ mi, $\mathbf{v}y = 18.7$ mi 4. $\mathbf{v}x = 338$ mi, $\mathbf{v}y = -294$ m

6. $\mathbf{v}x = -3590$ ft, $\mathbf{v}y = -46\bar{0}$ ft 8. 3.83 ft at 123.3° 10. 933 m at 172.3°

12. 572 mi/h at 41.7° 14. 96 Ω, 28° 16. $|\mathbf{R}| = 72\ \Omega$, $|\mathbf{X}_C| = 83\ \Omega$ 18. 124 kw/h at 222.9°

20. 28.1 km at 283.0° 22. 24.9 mi at 233.8° 24. 10,200 ft at 300.9° 26. 146 km at 120.9°

28. 69.3 mi at 161.1° (18.9° north of west)

Section 11.8

2. 32.7 km at 264.7° (5.3° west of south) 4. 1.41 km at 315.0° (45.0° south of east)

6. 5.0 mi 8. 153.6° 10. 0 12. (a) 104 mi/h (b) 146 mi/h

14. 155 mi/h at 110.8° (20.8° west of north) 16. 195 mi/h at 341.0° (19.0° south of east)

18. (a) 18.4° upstream (b) 38.1° upstream (c) 1.5° downstream 20. 25.0 lb, 43.3 lb

22. 887 N at 258.2° (78.2° south of west) 24. 197 lb at 83.6° (83.6° north of east)

26. 71.3 lb at 47.5° from **R** 28. 608 lb at 98.3° 30. 191 lb at 115.7° 32. 325 lb 34. 49.3°

CHAPTER 12

Section 12.1

2. $y = 5 \sin x$

   $5; 2\pi$

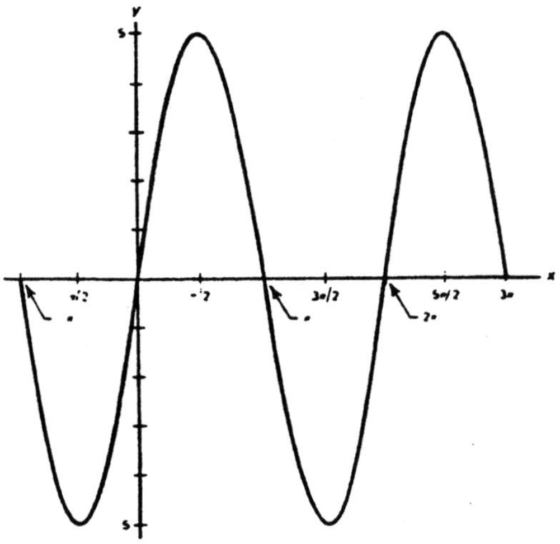

4. $y = -4 \cos x$

   $4; 2\pi$

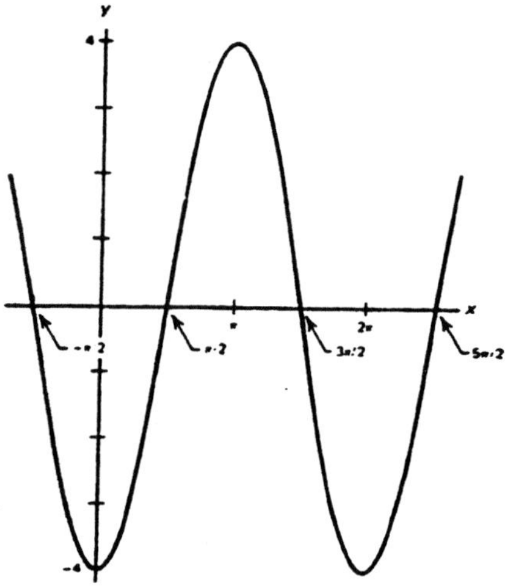

6. $y = \sin 4x$

   $1; \pi/2$

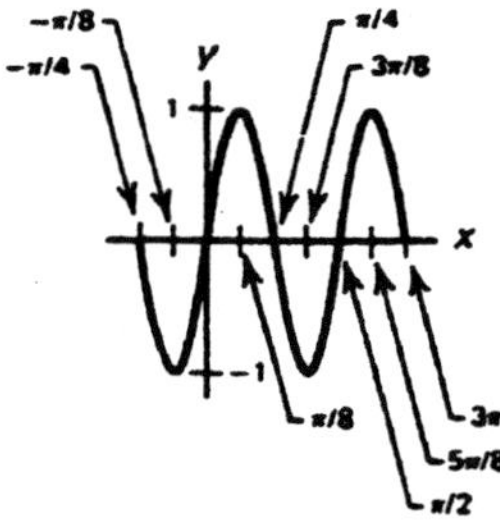

8. $y = \cos 3x$

   $1; 2\pi/3$

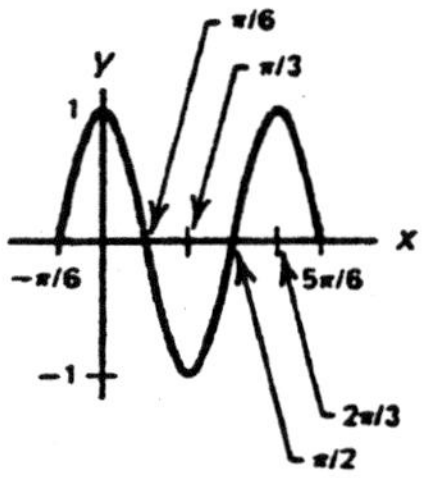

10. $y = 3 \cos 6x$

    $3; \pi/3$

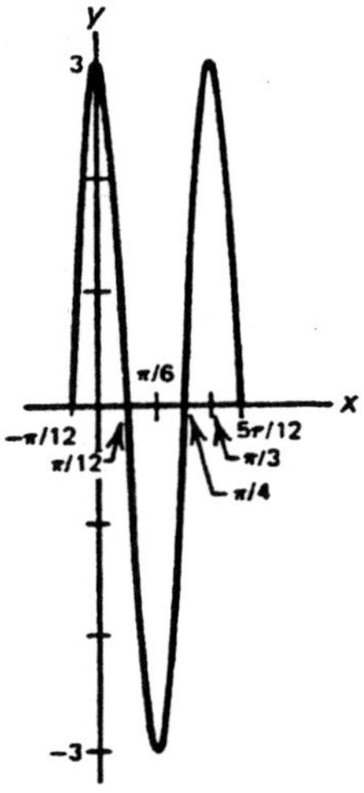

12. $y = \frac{3}{2} \sin \frac{3}{2} x$

    $\frac{3}{2}; \frac{4\pi}{3}$

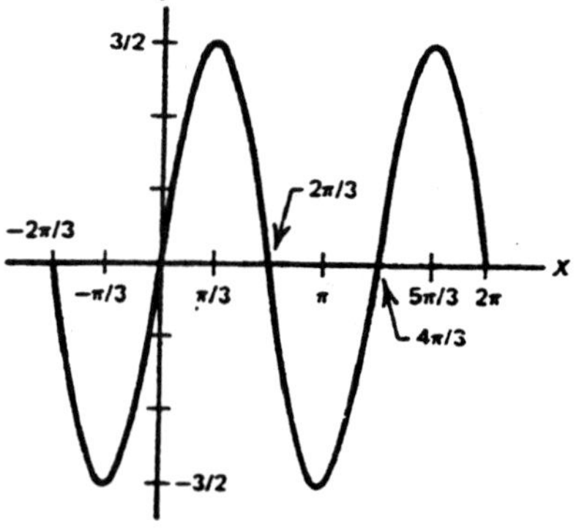

14. $y = -\dfrac{5}{2}\cos\dfrac{3}{4}x$

$5/2;\ 8\pi/3$

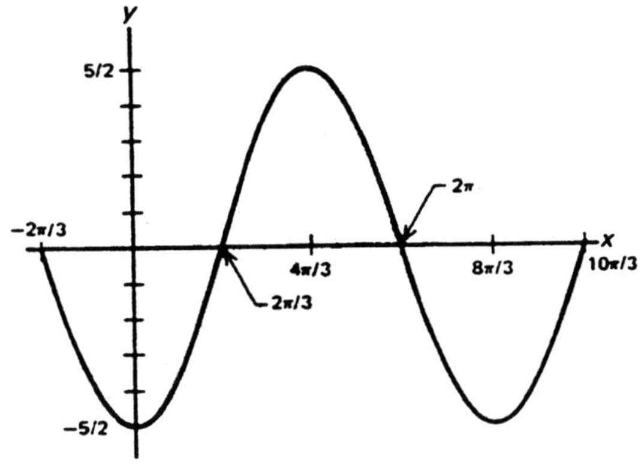

16. $y = 6\cos\dfrac{4\pi x}{3}$

$6;\ 3/2$

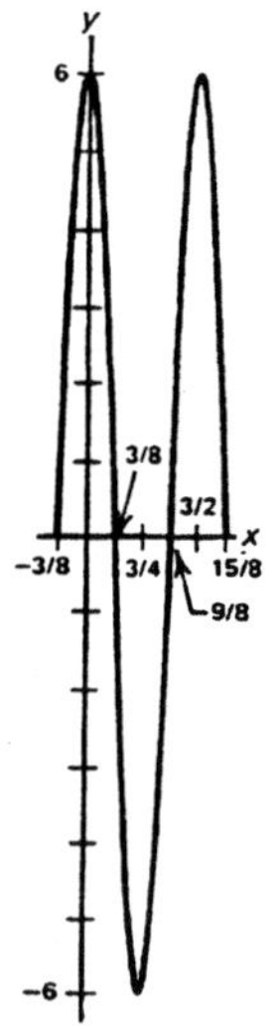

18. $y = -\sin\dfrac{\pi x}{2}$

$1;\ 4$

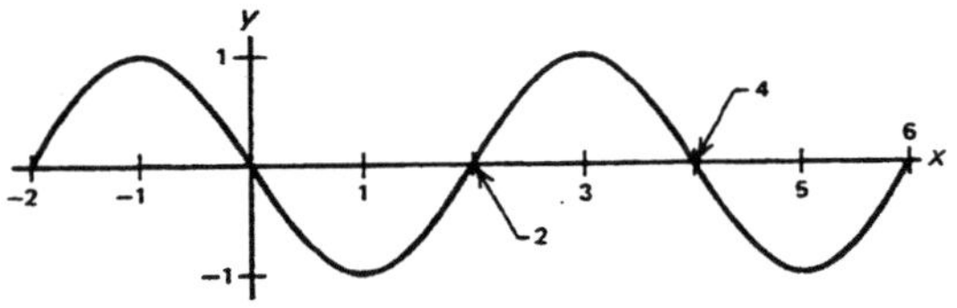

20. $y = 12\cos 160\pi x$

$12;\ \dfrac{1}{80}$

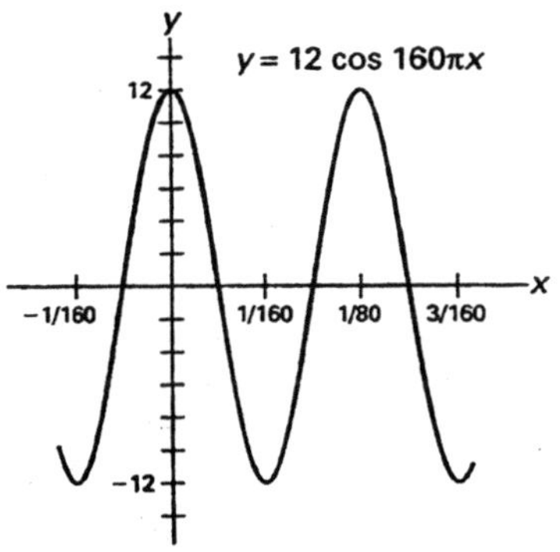

22. $y = 60\sin 40x$

$60;\ \dfrac{\pi}{20}$

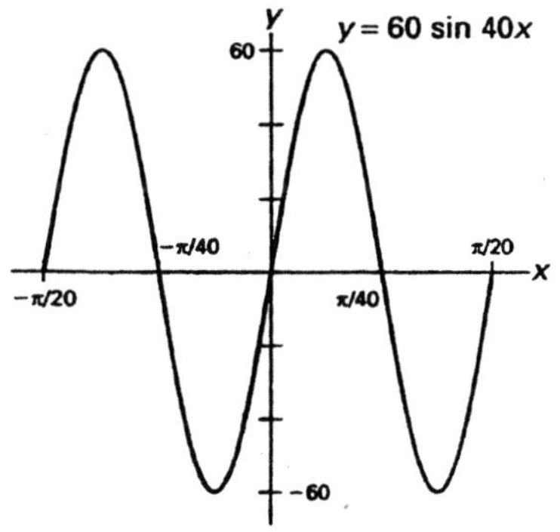

24. $y = -240\cos 120x$

$240;\ \dfrac{\pi}{60}$

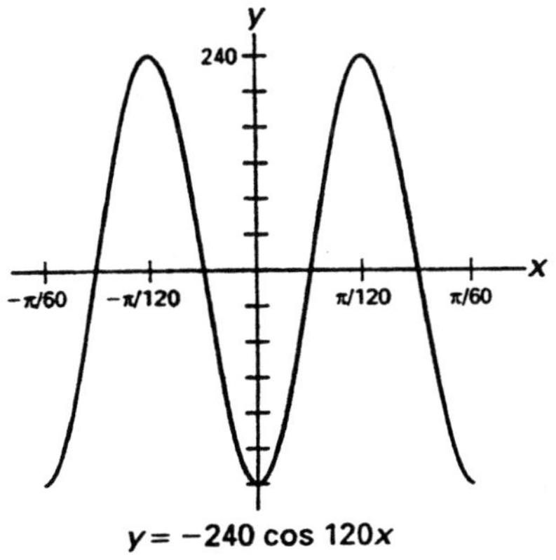

Section 12.2

2. $y = \cos\left(x + \frac{\pi}{4}\right)$

   1; 2π, $\pi/4$

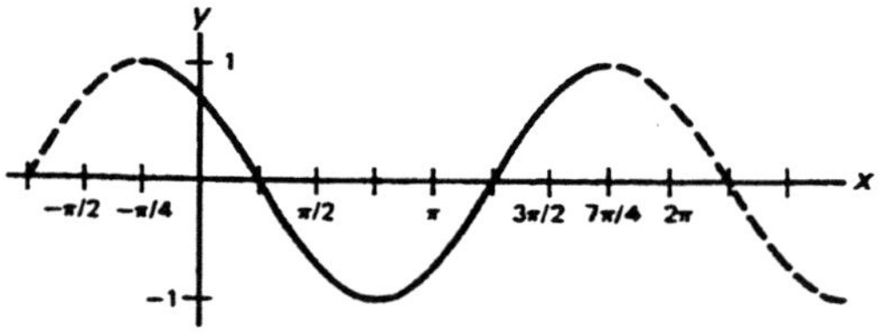

4. $y = 3\sin\left(x - \frac{\pi}{3}\right)$

   3; 2π; $-\pi/3$

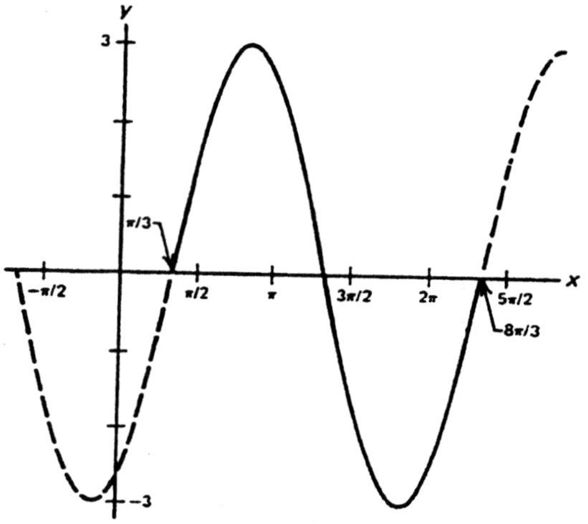

6. $y = \cos\left(2x + \frac{\pi}{3}\right)$

   1; π; $\pi/6$

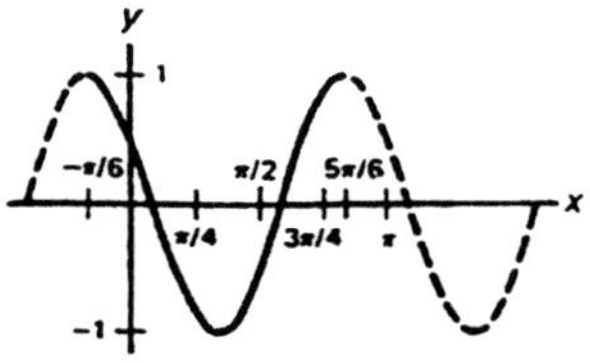

8. $y = -\sin\left(4x - \frac{2\pi}{3}\right)$

   1; $\pi/2$; $-\pi/6$

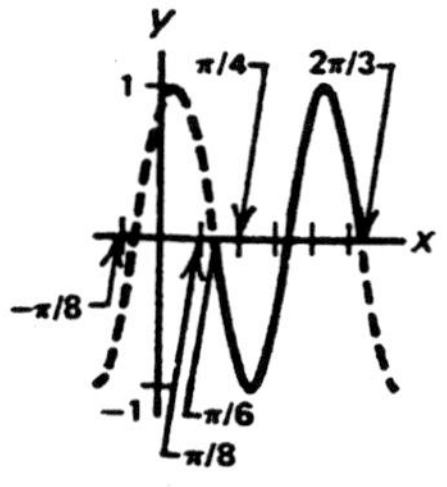

10. $y = 4\sin\left(\frac{1}{2}x - \frac{4\pi}{3}\right)$

    4; 4π, $-8\pi/3$

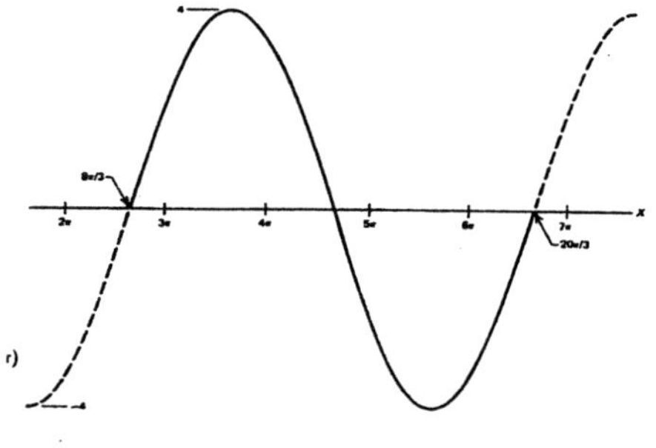

12. $y = 5\cos\left(\frac{2}{3}x + \pi\right)$

5; 3π; $3\pi/2$

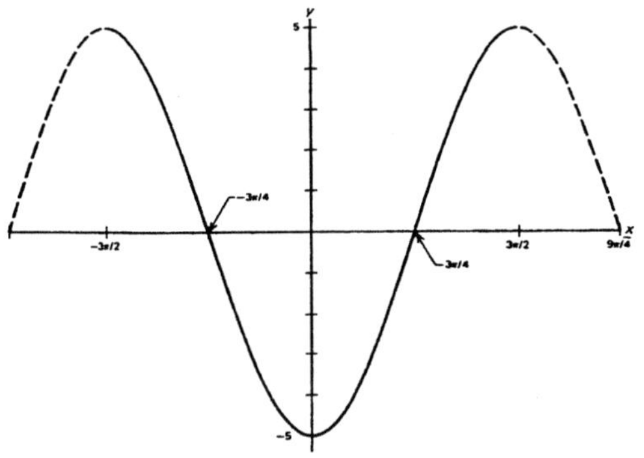

14. $y = 3\cos\left(\frac{\pi x}{4} + \frac{\pi}{8}\right)$

3; 8; 1/2

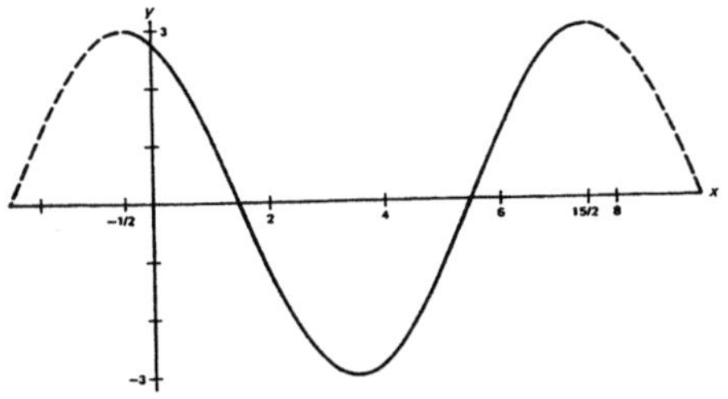

16. $y = 4\sin\left(\frac{\pi x}{6} - \frac{\pi}{3}\right)$

4; 12; −2

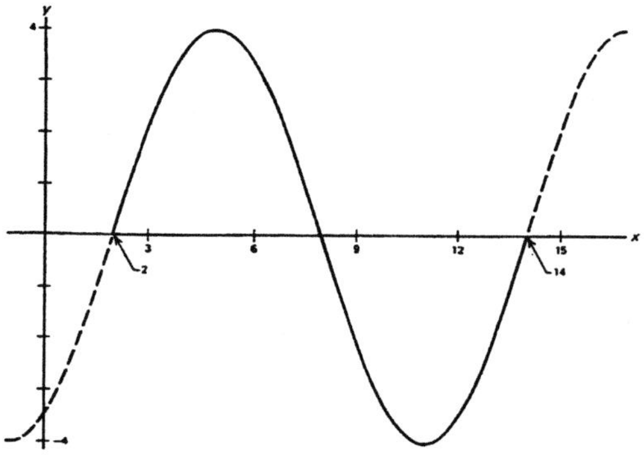

18. $y = 80\sin\left(40x - \frac{\pi}{4}\right)$

80; $\frac{\pi}{20}$; $\frac{\pi}{160}$

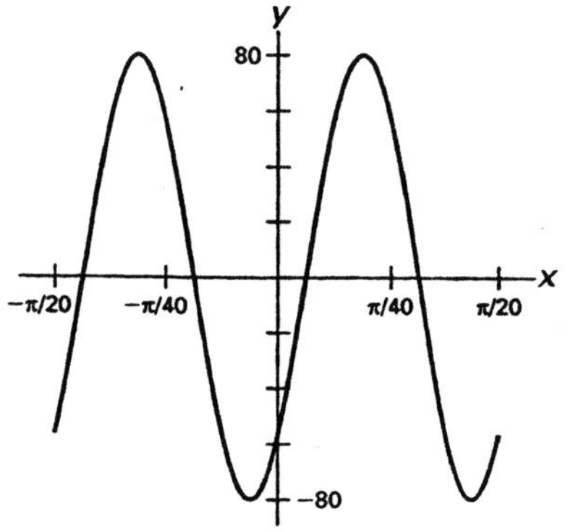

y = 80 sin (40x − π/4)

20. $y = 40\sin\left(20\pi x + \frac{\pi}{3}\right)$

40; $\frac{1}{10}$; $\frac{1}{60}$

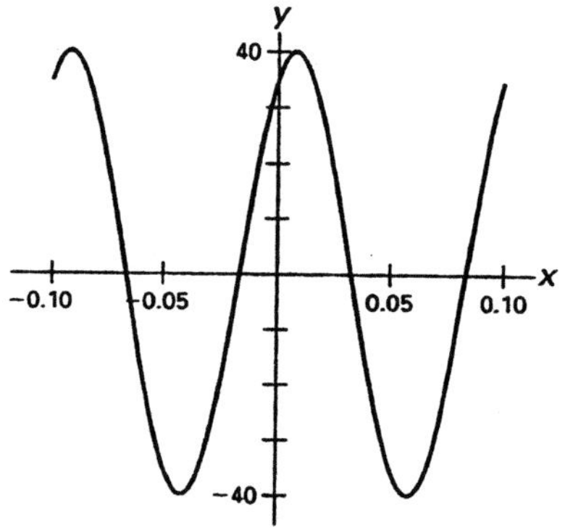

y = 40 sin (20πx + π/3)

22. $y = 4.5\cos\left(180x - \dfrac{\pi}{3}\right)$

$4.5;\ \dfrac{\pi}{90};\ \dfrac{\pi}{540}$

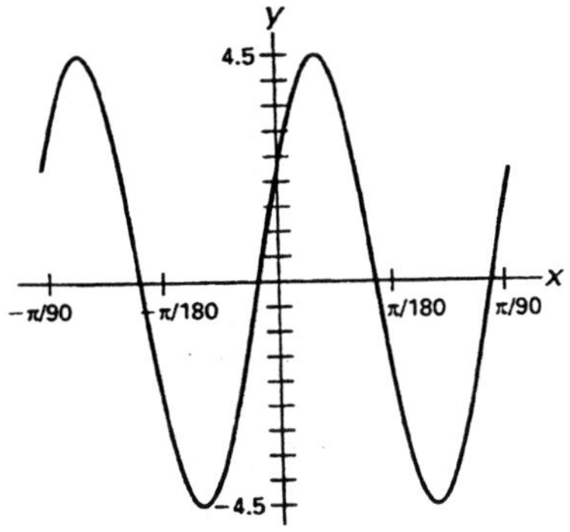

y = 4.5 cos (180x − π/3)

24. $y = 30\cos(160\pi x + 8\pi)$

$30;\ \dfrac{1}{80};\ \dfrac{1}{20}$

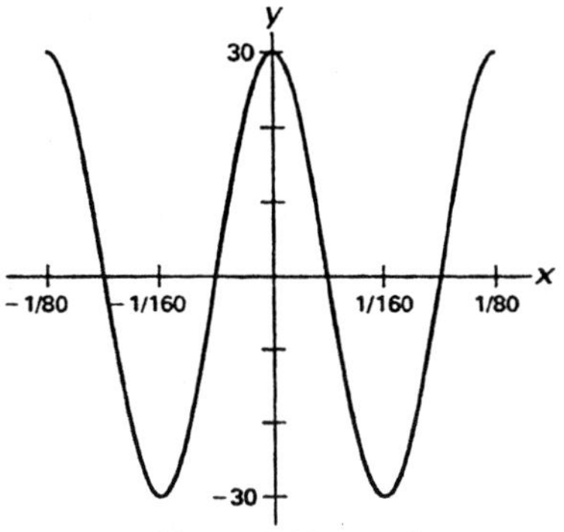

y = 30 cos (160πx + 8π)

Section 12.3

2. $y = \tan 4x$

$x/4$

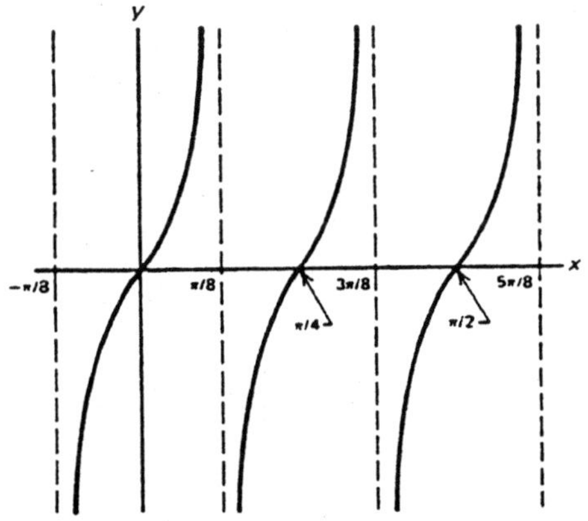

4. $y = -\tan\dfrac{3}{2}x$

$2\pi/3$

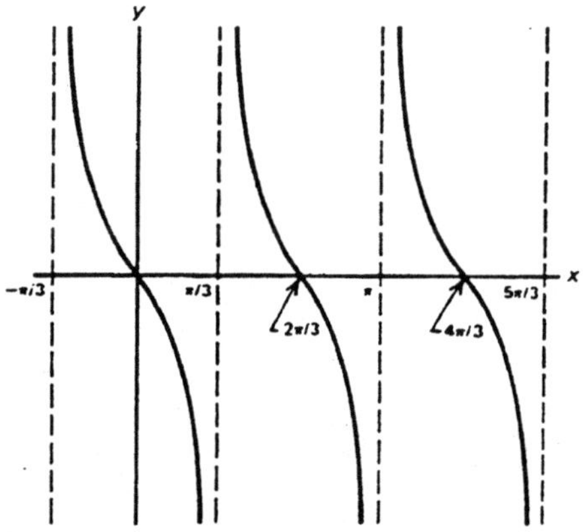

6. $y = 2\cot\dfrac{1}{3}x$

3π

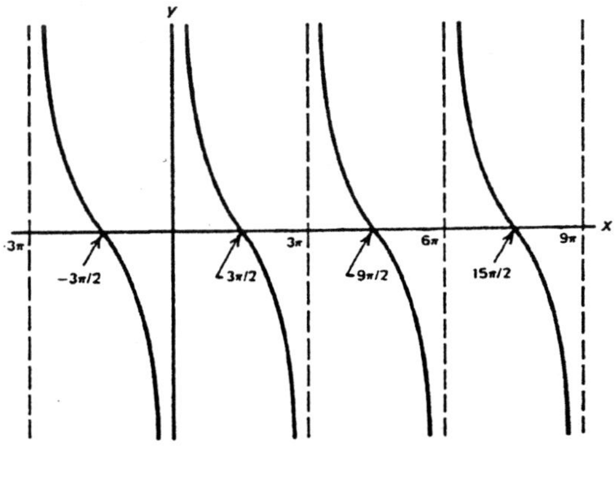

8. $y = 4\csc x$

2π

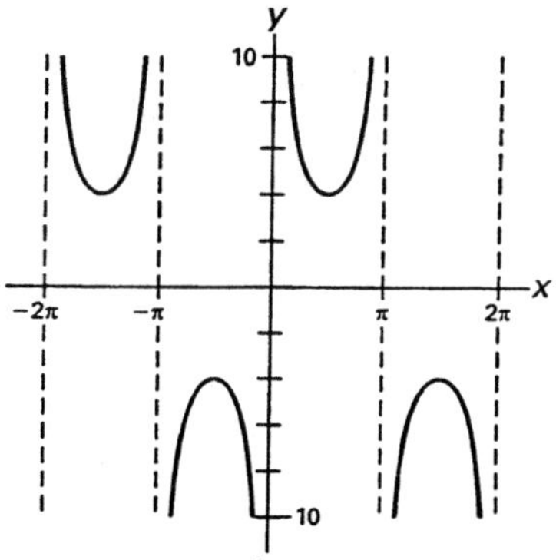

y = 4 cscx

10. $y = -5 \sec x$

2π

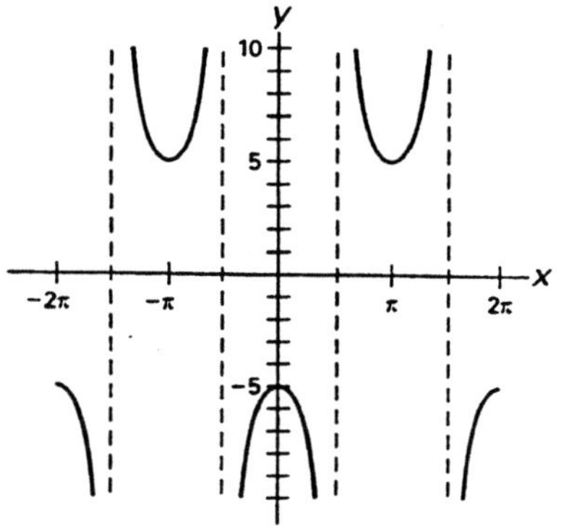

$y = -5 \sec x$

12. $y = 4 \sec 2x$; π

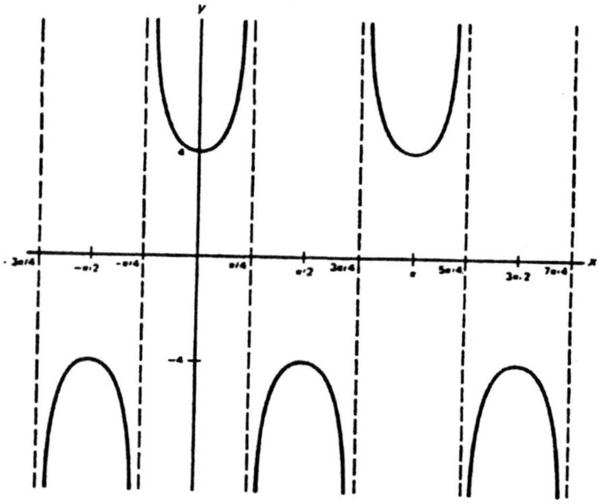

14. $y = -5 \sec \dfrac{3x}{2}$

$\dfrac{4\pi}{3}$

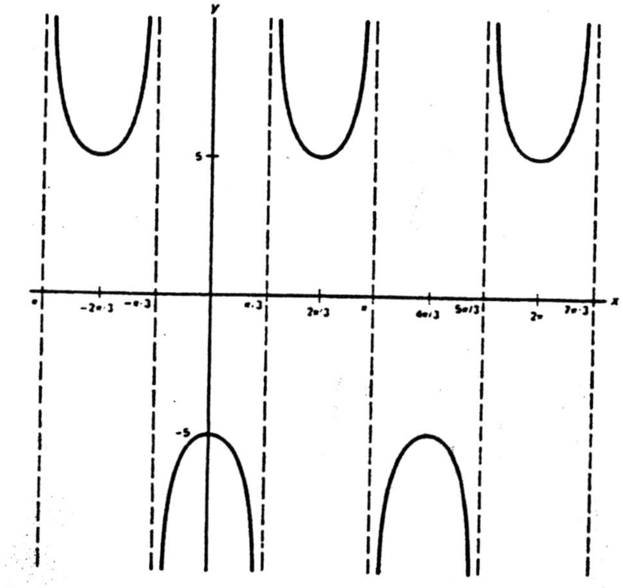

16. $y = 2 \csc 3x$

$2\pi/3$

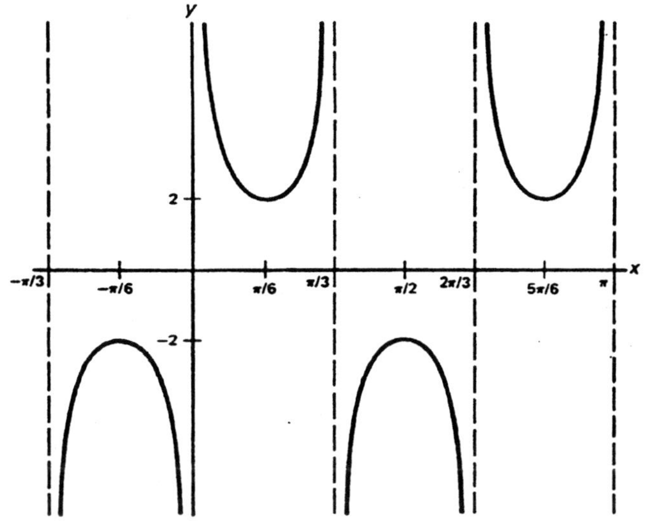

18. $y = \cot\left(2x - \frac{\pi}{2}\right)$

$\pi/2$

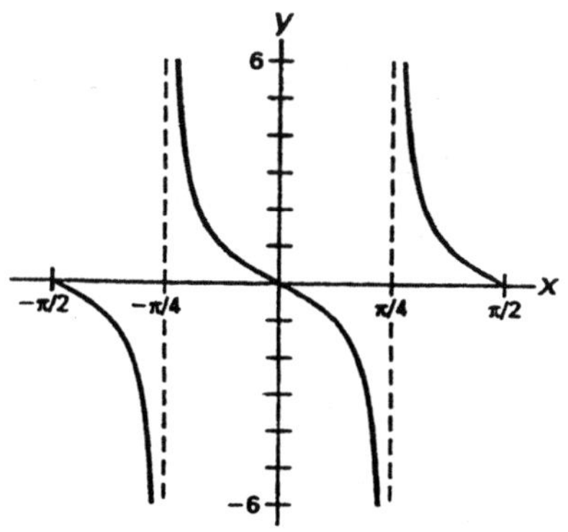

$y = \cot(2x - \pi/2)$

20. $y = 2 \csc\left(x + \frac{\pi}{3}\right)$

2π

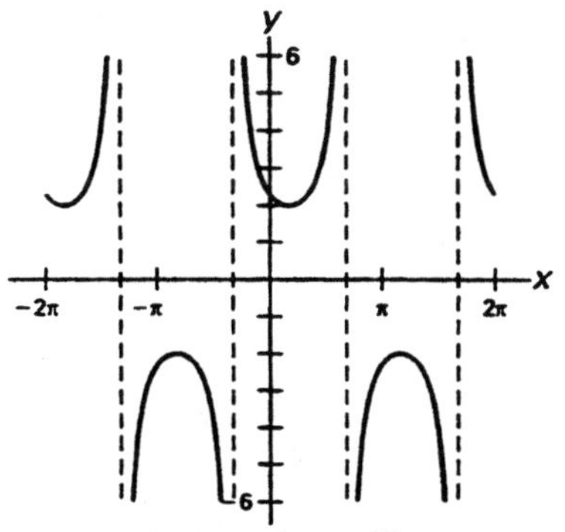

$y = 2 \csc(x + \pi/3)$

Section 12.4

2. $y = 2\cos x + \sin x$

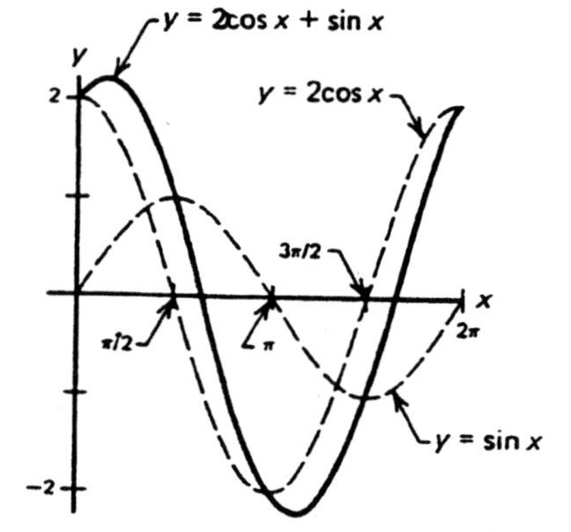

4. $y = \sin x + 3\cos x$

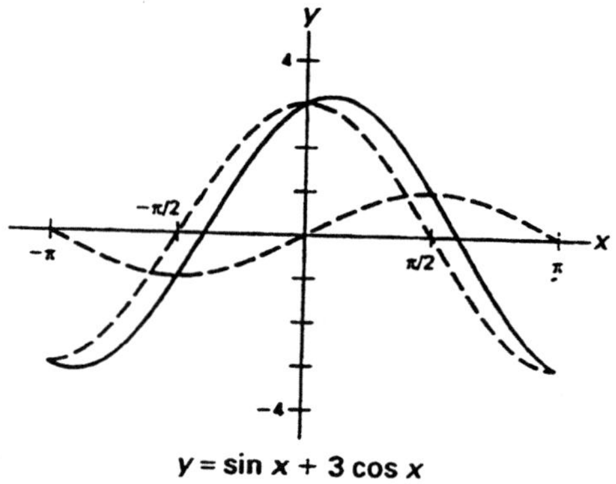

y = sin x + 3 cos x

6. $y = 2\cos(x/2) + \sin x$

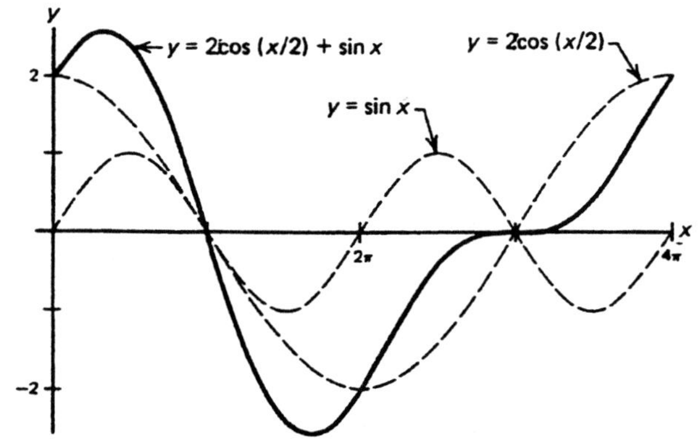

8. $y = 2\sin 2x + 2\cos x$

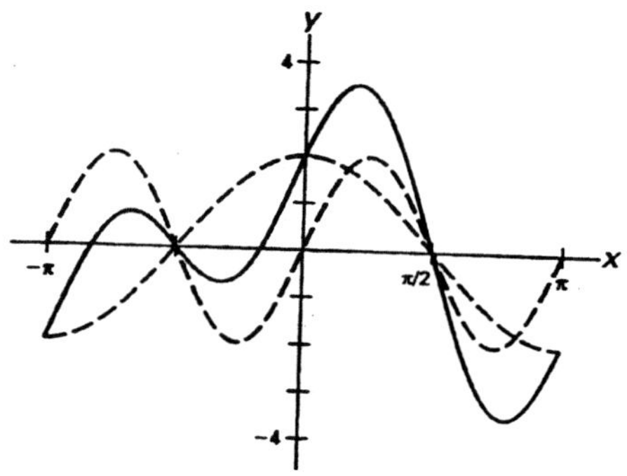

y = 2 sin 2 x + 2 cos x

10. $y = \cos x - 2\sin x$

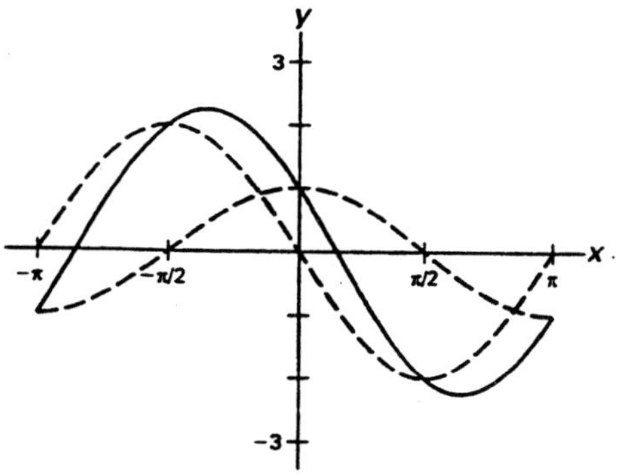

y = cos x − 2 sin x

12. $y = \cos 2x - \cos x$

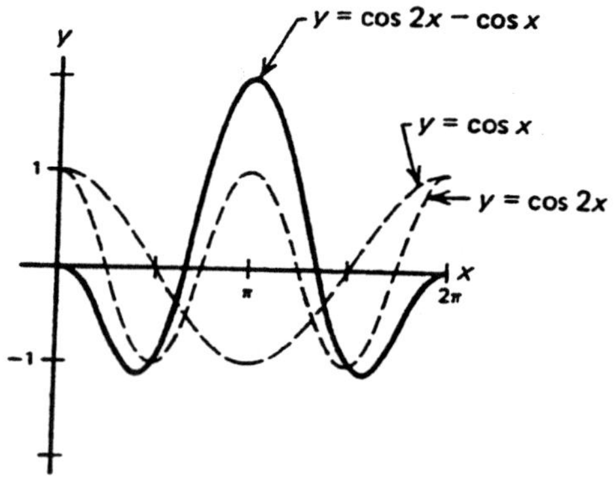

14. $y = x - 2\cos x$

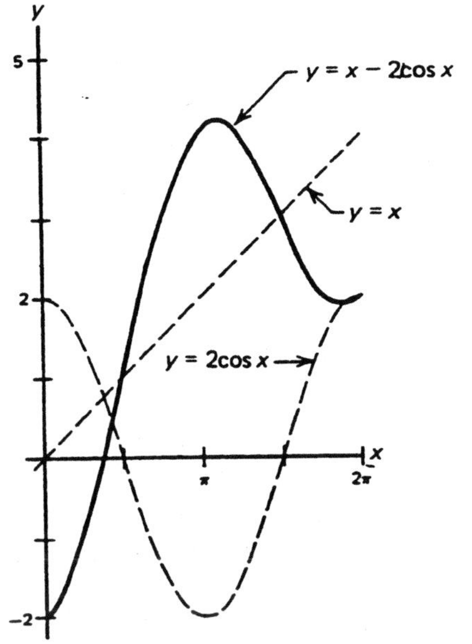

16. $y = \sin 2x + \cos(2x - \pi/4)$

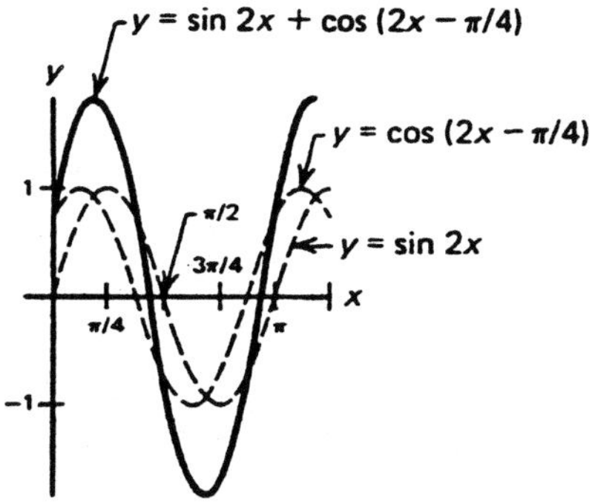

Section 12.5

2. $y = 8\sin(10\pi t + \pi/2)$
 or $y = 8\cos 10\pi t$

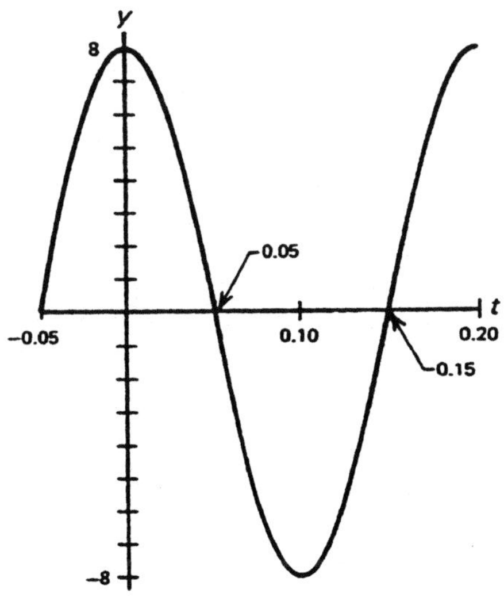

4. $y = 15\sin 120\pi t$

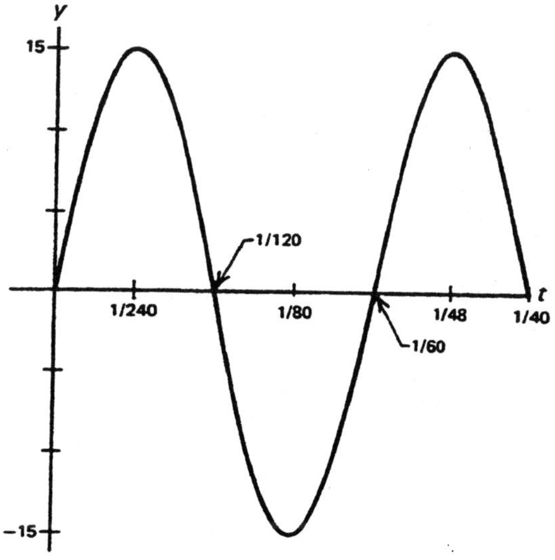

6. $e = 110 \cos(120\pi t - \pi/3)$

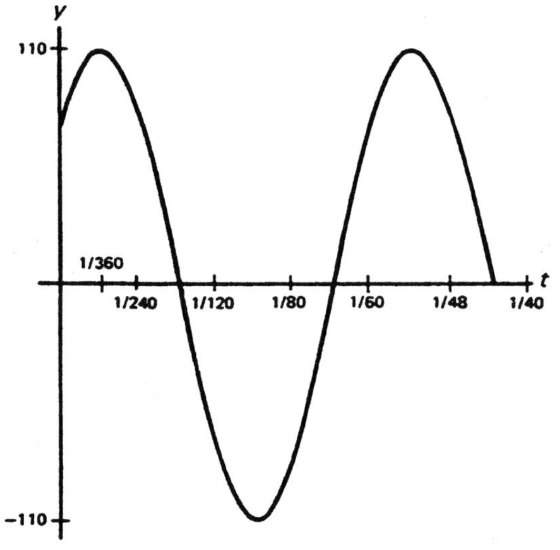

8. $y = 3 \sin 8\pi t$

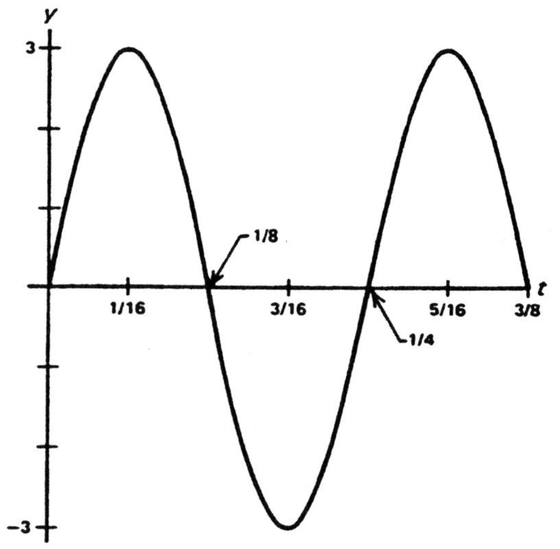

10. $5m$

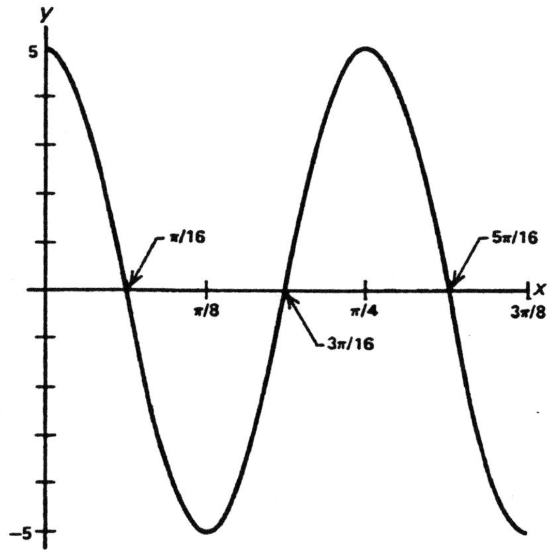

CHAPTER 13

Section 13.2

22. $\sin 3C$ 24. $\cos 3\theta$ 26. $-\sin\theta$ 28. $-\tan\theta$

30. $2\cos\theta\cos\phi$ 32. $\cos A$

Section 13.3

2. $5\sin^2 2x$ 4. $\sin 3\theta$ 6. 1 8. $10\cos 88°$ or 0.3490 10. $\sin 6x$

12. $\sin 8\theta$ 14. $\cos 14t$ 16. $\frac{15}{2}\sin\frac{x}{3}$ 18. $\frac{7}{25}$ 20. $\frac{\sqrt{26}}{26}$ 22. $-\frac{1}{3}$

Section 13.4

2. 240°, 300° 4. 45°, 135°, 225°, 315° 6. 60°, 120°, 240°, 300°

8. 40°, 80°, 160°, 200°, 280°, 320°

10. 30°, 60°, 120°, 150°, 210°, 240°, 300°, 330° 12. 30°, 150°, 270°

14. 90°, 135°, 270°, 315°

16. 67.5°, 292.5° 18. $\frac{\pi}{2}, \frac{7\pi}{6}, \frac{11\pi}{6}$ 20. $0, \pi, \frac{3\pi}{4}, \frac{7\pi}{4}$ 22. $\frac{\pi}{3}, \frac{2\pi}{3}, \frac{4\pi}{3}, \frac{5\pi}{3}$

24. $\frac{\pi}{6}, \frac{5\pi}{6}, \frac{7\pi}{6}, \frac{11\pi}{6}, \frac{\pi}{2}, \frac{3\pi}{2}$ 26. $0, \pi, \frac{\pi}{3}, \frac{5\pi}{3}$ 28. $0, \frac{\pi}{3}, \frac{2\pi}{3}, \pi, \frac{4\pi}{3}, \frac{5\pi}{3}$

30. $\frac{\pi}{3}, \pi, \frac{5\pi}{3}$ 32. $0, \frac{\pi}{3}, \frac{5\pi}{3}$ 34. 0

Section 13.5

2. y is the angle whose secant is x. 4. y is the angle whose cosine is $2x$.

6. y is one-half of the angle whose tangent is $3x$.

8. $x = \frac{1}{4}\arctan y$ 10. $x = \operatorname{arcsec}\frac{y}{3}$ 12. $x = \frac{1}{3}\arccos 2y$ 14. $x = \frac{3}{2}\arcsin\frac{2y}{5}$

16. $x = -\frac{1}{2} + \frac{1}{2}\arctan\frac{y}{4}$ 18. $x = \frac{1}{4} - \frac{1}{4}\operatorname{arcsec} 3y$

20. $\frac{\pi}{3}$ 22. $-\frac{\pi}{6}$ 24. $\frac{\pi}{6}$ 26. $-\frac{\pi}{2}$ 28. $\frac{\pi}{2}$ 30. $\frac{\pi}{2}$

32. $-\frac{\pi}{3}$ 34. 0.4115 36. 0.8481 38. 2.153 40. 1

42. $-\frac{1}{\sqrt{2}}$ 44. –2 46. $-\sqrt{3}$ 48. 2.4 50. 0.5218 52. $\frac{\sqrt{1-x^2}}{x}$

54. $\frac{1}{\sqrt{x^2-1}}$ 56. $\frac{x}{\sqrt{x^2+1}}$ 58. x 60. $\frac{\sqrt{1-9x^2}}{3x}$ 62. $\frac{1-x^2}{1+x^2}$

64.

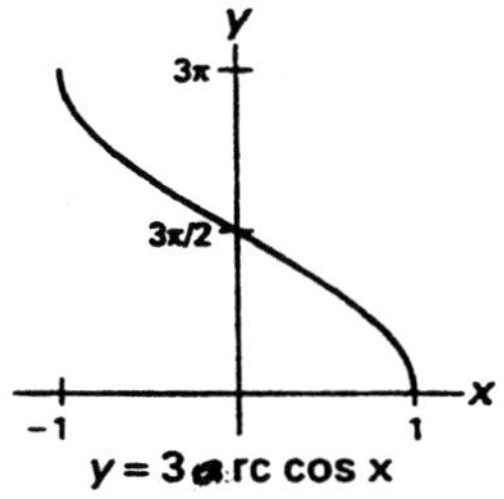

CHAPTER 14

Section 14.1

2. $10j$ 4. $11j$ 6. $6j\sqrt{2}$ 8. $10j\sqrt{5}$ 10. -1 12. $-j$ 14. j

16. 1 18. $4-2j$ 20. $-15+5j$ 22. 5 24. $24-62j$ 26. $-23-11j$

28. 40 30. $16-30j$ 32. $-1+j$ 34. $\frac{14-5j}{13}$ 36. $\frac{7+24j}{25}$

38. $\pm 2j\sqrt{3}$ 40. $\frac{1\pm j\sqrt{47}}{2}$ 42. $\frac{-1\pm j\sqrt{3}}{2}$ 44. $-1\pm 2j$ 46. $1, \frac{-1\pm j\sqrt{3}}{2}$

48. $0, \pm 3j$ 50. $0, \pm 5j$ 52. $0, 2, -1\pm j\sqrt{3}$ 54. $0, -4, 2\pm 2j\sqrt{3}$

56 through 60

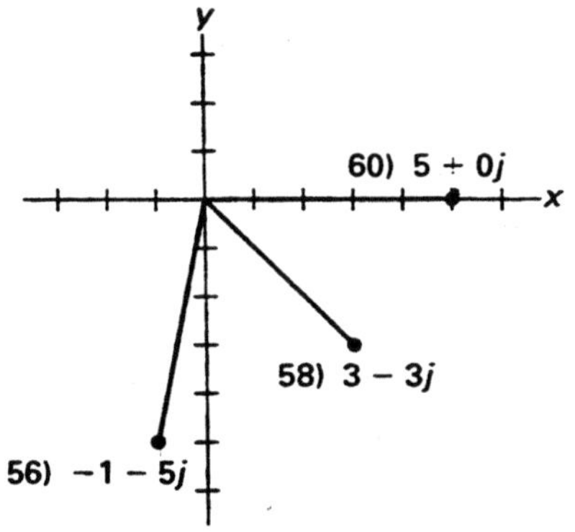

62. $-7-j$ 64. $-11-6j$ 66. $8-3j$ 68. 4

Section 14.2

2. $4\sqrt{2}(\cos 315° + j\sin 315°)$ 4. $8(\cos 150° + j\sin 150°)$ 6. $3(\cos 180° + j\sin 180°)$

8. $13(\cos 67° + j\sin 67°)$ 10. $\sqrt{34}(\cos 329° + j\sin 329°)$ 12. $-3\sqrt{3}+3j$

14. $-\frac{3}{2}-\frac{3\sqrt{3}}{2}j$ 16. $-\frac{\sqrt{6}}{2}-\frac{\sqrt{6}}{2}j$ 18. $1.69+3.63j$ 20. $-2.96-5.94j$

22. $3\sqrt{2}\,e^{2.36j}$ 24. $4e^{1.05j}$ 26. $\sqrt{74}\,e^{5.33j}$

28. $5(\cos 173° + j\sin 173°), -4.96+0.61j$ 30. $1(\cos 396° + j\sin 396°), 0.81+0.59j$

32. $6(\cos 180° + j\sin 180°), -6+0j$

Section 14.3

2. $48e^{5j}$ 4. $24e^{8j}$ or $24e^{1.72j}$ 6. $21e^{-2.7j}$ or $21e^{3.6j}$ 8. $-13.0+7.50j$

10. $3.92-44.8j$ 12. $-2.96-0.469j$ 14. $10(\cos 298° + j \sin 298°)$

16. $18(\cos 180° + j \sin 180°)$ 18. $72(\cos 313° + j \sin 313°)$

20. $-2\sqrt{2}-2j\sqrt{2}$ or $-2.83-2.83j$ 22. $-27j$ 24. $-19.7+7.18j$ 26. $1.5e^{2j}$

28. 12^{2j} 30. $4e^{5.7j}$ 32. $\frac{2}{3}$ 34. $6.8+5.9j$ 36. $-0.73+1.0j$

38. $9(\cos 100° + \text{j} \sin 100°)$ 40. $3.5(\cos 67° + j \sin 67°)$ 42. $6(\cos 150° + j \sin 150°)$

44. $-\frac{\sqrt{2}}{8}+\frac{\sqrt{2}}{8}j$ or $-0.177+0.177j$ 46. $-5j$ 48. $\frac{16}{5}j$ 50. $27j$ 52. -16

Section 14.4

2. $8e^{1.97j}$ 4. $64e^{3.9j}$ 6. $1(\cos 175° + j \sin 175°)$ 8. $125(\cos 105° + j \sin 105°)$

10. $1(\cos 200° + j \sin 200°)$ 12. $-1+j\sqrt{3}$ 14. $512j$ 16. $3110+272j$ 18. $\frac{1}{8}$

20. $0.924+0.383j$, $-0.924+0.383j$, $-0.924-0.383j$; $0.924-0.383j$

22. $\frac{\sqrt{3}}{2}+\frac{j}{2}, j, -\frac{\sqrt{3}}{2}+\frac{j}{2}, -\frac{\sqrt{3}}{2}-\frac{j}{2}, -j, \frac{\sqrt{3}}{2}-\frac{j}{2}$

24. $3(\cos 45° + j \sin 45°)$, $\frac{3\sqrt{2}}{2}+\frac{3\sqrt{2}}{2}j$

$3(\cos 135° + j \sin 135°)$, $-\frac{3\sqrt{2}}{2}+\frac{3\sqrt{2}}{2}j$

$3(\cos 225° + j \sin 225°)$, $-\frac{3\sqrt{2}}{2}-\frac{3\sqrt{2}}{2}j$

$3(\cos 315° + j \sin 315°)$, $\frac{3\sqrt{2}}{2}-\frac{3\sqrt{2}}{2}j$

26. $1(\cos 60° + j \sin 60°)$, $\frac{1}{2}+\frac{\sqrt{3}}{2}j$

$1(\cos 180° + j \sin 180°)$, -1

$1(\cos 300° + j \sin 300°)$, $\frac{1}{2}-\frac{\sqrt{3}}{2}j$

28. $1(\cos 90° + j \sin 90°), j$

$1(\cos 210° + j \sin 210°), -\frac{\sqrt{3}}{2} - \frac{j}{2}$

$1(\cos 330° + j \sin 330°), \frac{\sqrt{3}}{2} - \frac{3\sqrt{2}}{2} j$

30. $1(\cos 15° + j \sin 15°)$, $0.966 + 0.259j$
$1(\cos 75° + j \sin 75°)$, $0.259 + 0.966j$
$1(\cos 135° + j \sin 135°)$, $-0.707 + 0.707j$
$1(\cos 195° + j \sin 195°)$, $-0.966 - 0.259j$
$1(\cos 255° + j \sin 255°)$, $-0.259 - 0.966j$
$1(\cos 315° + j \sin 315°)$, $0.707 - 0.707j$

32. $2(\cos 0° + j \sin 0°)$, 2
$2(\cos 120° + j \sin 120°)$, $-1 + \sqrt{3}j$
$2(\cos 240° + j \sin 240°)$, $-1 - \sqrt{3}j$

34. $2^{1/4}(\cos 23° + j \sin 23°)$, $1.09 + 0.465j$
$2^{1/4}(\cos 203° + j \sin 203°)$, $-1.09 - 0.465j$

CHAPTER 15

Section 15.1

2. 3×2 4. 1×5 6. 3×4 8. 2×3 10. $x = 9, y = 3, z = 7$

12. $x = 4, y = 8, a = -8, b = -6$ 14. $x = 6, y = 8, z = 2$

16. $x = 9, y = 5, z = 6, a = 8, b = 3, c = 7$

18. $\begin{bmatrix} 5 & -17 \\ 4 & 5 \\ 3 & 9 \end{bmatrix}$ 20. $\begin{bmatrix} 7 & -4 & 10 \end{bmatrix}$ 22. $\begin{bmatrix} -1 & 9 \\ -4 & -3 \end{bmatrix}$ 24. $\begin{bmatrix} 2 & -18 \\ 8 & 6 \end{bmatrix}$

26. $\begin{bmatrix} 0 & 10 & 40 \\ -20 & -35 & -25 \end{bmatrix}$ 28. $\begin{bmatrix} 1 & 12 & 14 \\ 1 & -9 & -5 \end{bmatrix}$ 30. Not defined 32. $\begin{bmatrix} 2 & 14 & -12 \\ 22 & 15 & 15 \end{bmatrix}$

34. $\begin{bmatrix} -2 & -30 & -52 \\ 10 & 39 & 25 \end{bmatrix}$ 36. Not defined 42. $\begin{bmatrix} 35 & 35 & 18 \\ 25 & 21 & 10 \\ 15 & 7 & 5 \end{bmatrix}$

44. $\begin{bmatrix} 3 & -5 & -2 & -1 \\ 5 & -1 & 2 & -8 \\ -3 & -13 & -8 & 0 \\ -5 & -2 & 2 & -1 \end{bmatrix}$

Section 15.2

2. $\begin{bmatrix} 18 \end{bmatrix}$ 4. $\begin{bmatrix} -17 & -3 \end{bmatrix}$ 6. $\begin{bmatrix} 2 & -10 \\ -2 & 19 \end{bmatrix}$ 8. $\begin{bmatrix} 4 & 6 & 8 & 10 \\ 1 & 3 & 5 & 8 \end{bmatrix}$

10. $\begin{bmatrix} 9 & 10 \\ -5 & -8 \end{bmatrix}$ 12. Not defined 14. $\begin{bmatrix} 15 & 9 \\ 9 & -5 \\ 19 & 27 \end{bmatrix}$ 16. $\begin{bmatrix} 12 & 59 & 31 \\ 2 & -59 & 15 \\ 16 & 46 & -18 \end{bmatrix}$

18. Not defined 20. $\begin{bmatrix} -34 \\ 56 \\ -20 \end{bmatrix}$ 22. $\begin{bmatrix} 39 & -4 \end{bmatrix}$ 24. (a) $\begin{bmatrix} 5 & 2 \\ 10 & 9 \end{bmatrix}$

24. (b) $\begin{bmatrix} -29 & 10 \\ -20 & -39 \end{bmatrix}$ 24. (c) $\begin{bmatrix} 1 & 8 \\ -16 & -7 \end{bmatrix}$ 26. (a) $\begin{bmatrix} 4 & -5 & 10 \\ -5 & -11 & 43 \\ 5 & -55 & 44 \end{bmatrix}$

26. (b) $\begin{bmatrix} 131 & 2 & 190 \\ 55 & 46 & 89 \\ 280 & -68 & 389 \end{bmatrix}$ 26. (c) $\begin{bmatrix} 4 & -5 & 10 \\ -5 & -11 & 43 \\ 5 & -55 & 44 \end{bmatrix}$ 28. (a) $\begin{bmatrix} 3 & 24 \\ 3 & -36 \\ 29 & 12 \end{bmatrix}$

28. (b) Not defined 28. (c) $\begin{bmatrix} 34 & 12 \\ 10 & 30 \end{bmatrix}$

Section 15.3

2. $\begin{bmatrix} \frac{3}{23} & \frac{-2}{23} \\ \frac{4}{23} & \frac{5}{23} \end{bmatrix}$ 4. $\begin{bmatrix} 1 & 1 \\ -3.5 & -3 \end{bmatrix}$ 6. $\begin{bmatrix} -0.2 & 0.2 \\ 0.4 & 0.6 \end{bmatrix}$ 8. $\begin{bmatrix} \frac{1}{7} & \frac{3}{7} \\ \frac{1}{14} & \frac{5}{14} \end{bmatrix}$

10. $\begin{bmatrix} -\frac{2}{3} & \frac{5}{6} \\ \frac{1}{3} & -\frac{1}{6} \end{bmatrix}$ 12. $\begin{bmatrix} -3 & -1 \\ -2 & -0.5 \end{bmatrix}$ 14. $\begin{bmatrix} 1 & 1.5 \\ -1 & -1.25 \end{bmatrix}$ 16. $\begin{bmatrix} -\frac{1}{3} & \frac{1}{9} \\ \frac{1}{3} & \frac{2}{9} \end{bmatrix}$

18. $\begin{bmatrix} 4.5 & -2.5 & 0.5 \\ -2.5 & 0.5 & 0.5 \\ 0.5 & 0.5 & -0.5 \end{bmatrix}$ 20. $\begin{bmatrix} \frac{1}{3} & \frac{1}{6} & -\frac{1}{3} \\ -\frac{1}{6} & \frac{5}{12} & -\frac{1}{3} \\ \frac{1}{3} & -\frac{1}{3} & \frac{2}{3} \end{bmatrix}$ 22. $\begin{bmatrix} \frac{1}{3} & \frac{1}{6} & 0 \\ -\frac{1}{3} & \frac{1}{3} & 0 \\ 0 & 0 & 1 \end{bmatrix}$

24. $\begin{bmatrix} -0.4 & 1 & 1.2 \\ -0.2 & 1 & 0.6 \\ 0.6 & -1 & -0.8 \end{bmatrix}$ 26. $\begin{bmatrix} \frac{6}{17} & -\frac{2}{17} & \frac{1}{17} & -\frac{6}{17} \\ \frac{2}{17} & \frac{5}{17} & \frac{6}{17} & -\frac{2}{17} \\ \frac{11}{68} & -\frac{15}{68} & -\frac{1}{68} & \frac{3}{34} \\ \frac{9}{34} & -\frac{3}{34} & -\frac{7}{34} & \frac{4}{17} \end{bmatrix}$ 28. $\begin{bmatrix} -2.25 & 1.25 & 0 & 0.5 \\ 3.25 & -1.75 & 0.25 & -0.5 \\ -0.5 & 0.5 & 0 & 0 \\ 4.25 & -2.25 & 0 & -0.5 \end{bmatrix}$

30. $x = -6, y = -3$ 32. $x = -5, y = 6$ 34. $x = 2, y = -2$ 36. $x = 3, y = 4$

38. $x = 1, y = 4, z = 7$ 40. $x = 5, y = -6, z = 2$

Section 15.4

2. $x = 0, y = -6$

4. $x = -6, y = 12, z = 40$

6. $x = -4, y = 6, z = 20, w = -12$

8. $x = 5, y = 7$

10. $x = -2, y = 9$

12. $x = 5, y = -6$

14. $x = \frac{-23}{13}, y = \frac{-128}{39}$

16. $x = 4, y = 3, z = -1$

18. $x = 3, y = -8, z = 2$

20. $x = 6, y = -1, z = 1$

22. $x = 4, y = 5, z = -2, w = 1$

24. $x = 2, y = -2, z = 3, w = 3$

26. $x = 230, y = 315, z = 290$

CHAPTER 16

Section 16.1

2. x: –4, –3, 2; y: –24 4. x: –3, 1, 1.25; y: –15 6. x: –2, –1, 2; y: –4

8. $f(x) = (x + 2)(x + 1)(x - 2)$ 10. $f(x) = (x + 3)(x + 2)(x - 4)$

12. $f(x) = x(x + 2)(x - 4)$ 14. $f(x) = (x + 3)(x - 1)^2$ 16. $f(x) = x(x - 3)^2$

18. $f(x) = (x + 4)(x + 1)(x - 2)(x - 3)$ 20. $f(x) = (x + 4)(x + 3)(x - 1)^2$

22. $f(x) = x^2(x + 2)(x - 3)$ 24. $f(x) = (x + 1)^2(x - 4)^2$

Section 16.2

2. est: –1, 1.8, 4; better: –0.930802, 1.83704, 4.09376

4. est: –0.22; 0.64, 1.9; better: –0.252567, 0.699705, 1.88620

6. est: 0.8; better: 0.794431

8. est: –1.2, 0.5, 1.0, 1.9; better: –1.24698, 0.445042, 1.00000, 1.80194

10. est: –2.1, 1.5; better: –2.21564, 1.50813

Section 16.3

2. 3 4. 3 6. 8 8. $-5 \pm j$, 2.54400 10. $\pm 4j$, $1 \pm 3j$

12. $1 \pm 2j$ (multiplicity two) 14. –2, 1, $1 \pm 3j$ 16. $f(x) = x^3 - 5x^2 + 17x - 13 = 0$

18. $f(x) = 8x^4 - 28x^3 + 26x^2 + 3x - 9 = 0$ 20. $f(x) = x^4 - 12x^3 + 37x^2 - 12x + 36 = 0$

22. $f(x) = 3x^6 - 11x^5 + 18x^4 - 20x^3 + 17x^2 - 9x + 2 = 0$

24. Since complex roots occur in conjugate pairs, there will be six roots for a fifth degree equation.

CHAPTER 17

Section 17.1

2. 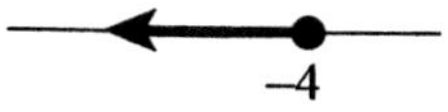

4. 

6. 

8. 

10. 

12. $x < 0$

14. $x \geq 8$ 16. $0 \leq x \leq 3$ 18. $1 < x \leq 4$ 20. $x \leq -3$ or $x \geq 0$

22. $x \geq 6$ 

24. $x > 4$ 

26. $x > -1$ 

28. $x \geq -6$ 

30. $x > -6$ 

32. $x < -\frac{15}{4}$ 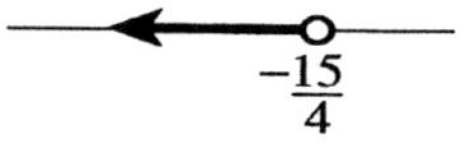

34. $x \leq -15$ 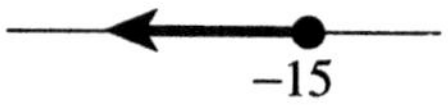

36. $x \geq 5$ 

38. $x < 4.2$ 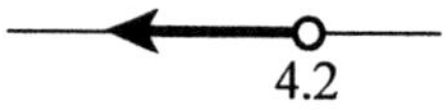

40. $x < 0.5$ 

42. 

44. 

46. 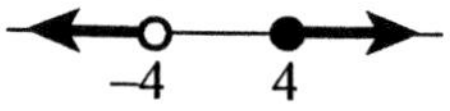

48. no graph

50. 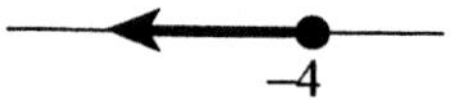

52. $4 < x < 8$ 

54. $-2 \leq x < 4$ 

56. $-3 \leq x < 0.5$ 

58. $-6 \leq x \leq 4$ 

Section 17.2

2. 12, –2 4. $-\frac{1}{4}, -\frac{5}{4}$ 6. $-2, -\frac{1}{3}$ 8. 0 10. $-\frac{4}{7}, -2$

12. $x < -4$ or $x > 4$ 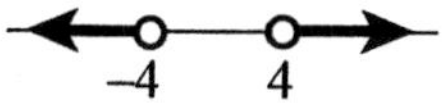

14. $-16 < x < 2$ 

16. $x < -\frac{2}{3}$ or $x > 2$ 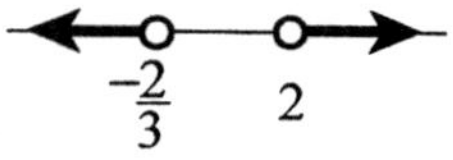

18. $-6 \leq x \leq 7$ 

20. $x \leq -15$ or $x \geq 8$ 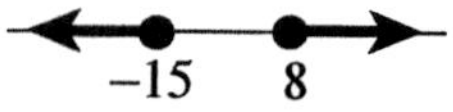

22. $x > \frac{7}{3}$ or $x < \frac{-5}{3}$

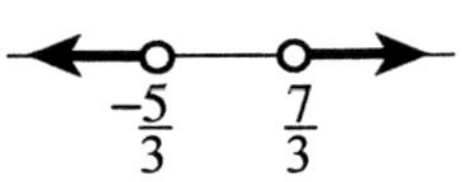

24. $-14.5 \le x \le 10$

26. $-5 < x < -1$

28. all real numbers except $-\frac{1}{3}$

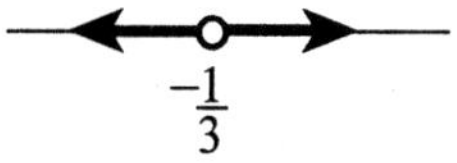

Section 17.3

2. $-7 < x < 2$ 4. $x \le \frac{1}{3}$ or $x \ge \frac{5}{6}$ 6. $-5 < x < 1$ 8. $\frac{2}{3} \le x \le 2$

10. $x < -\frac{1}{2}$ or $x > 3$ 12. $x \le \frac{1}{3}$ or $x \ge 4$ 14. $x = \frac{5}{3}$ 16. $-3 \le x \le 3$

18. $x < -2\sqrt{3}$ or $x > 2\sqrt{3}$ 20. $0 \le x \le 4$ 22. $x < -5$ or $-4 < x < 0$

24. $x \le 1$ 26. $x \le -3$ or $-1 \le x \le 1$ or $x \ge 3$ 28. $x \le -7$ or $x > -3$

30. $x < -4$ or $1 \le x < 2$ 32. $-3 < x < -2$ or $2 \le x \le 3$ 34. $x < -3$ or $3 < x < 4$ or $x > 6$

36. $x = -1$ or $x = 0$ or $1 \le x < 4$ or $x > 5$ 38. $x < 4$ or $x \ge 5$ 40. $-3 < x \le -2.8$

Section 17.4

2.

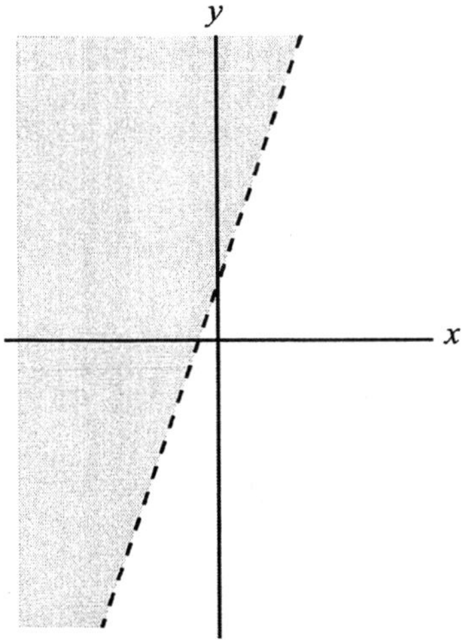

$y > 3x + 1$

4.

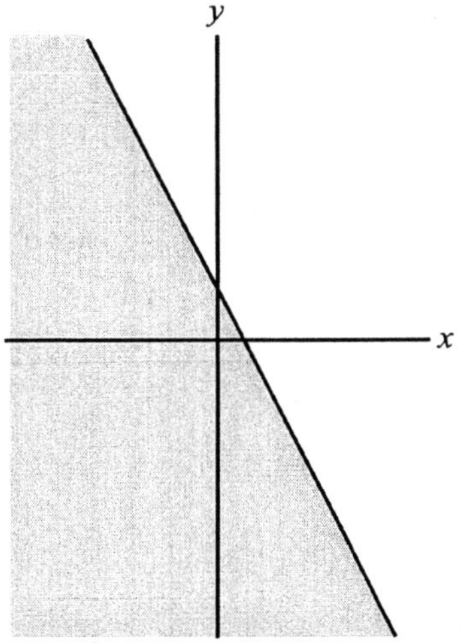

$y \le 1 - 2x$

6.

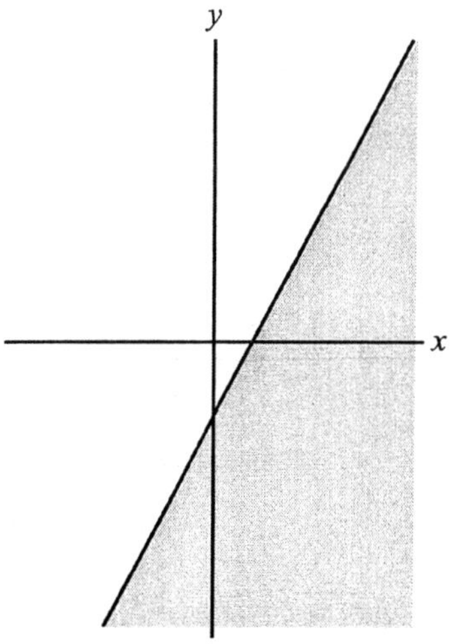

$6x - 4y \ge 7$

8.

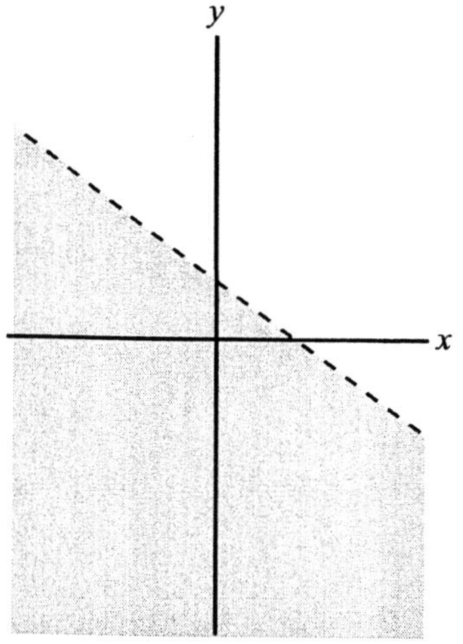

$3x + 5y < 4$

10.

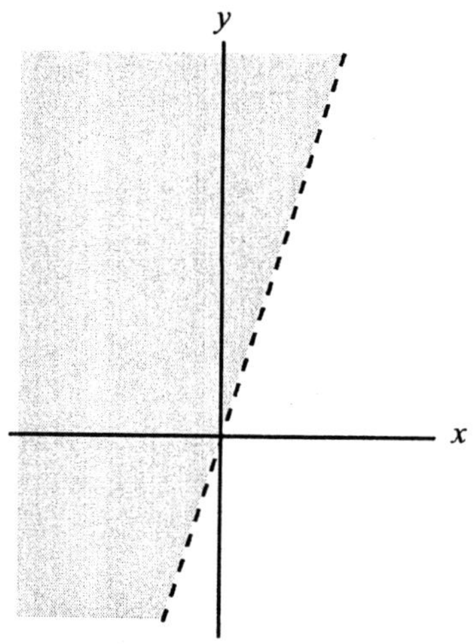

$y > 3x$

12.

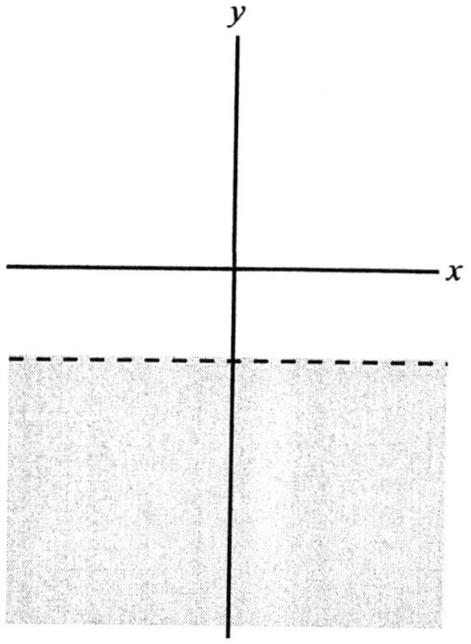

$y < -2$

14.

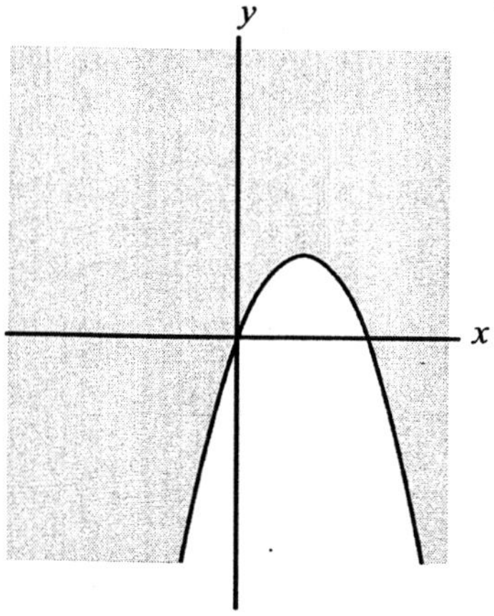

$y \geq 2x - x^2$

16.

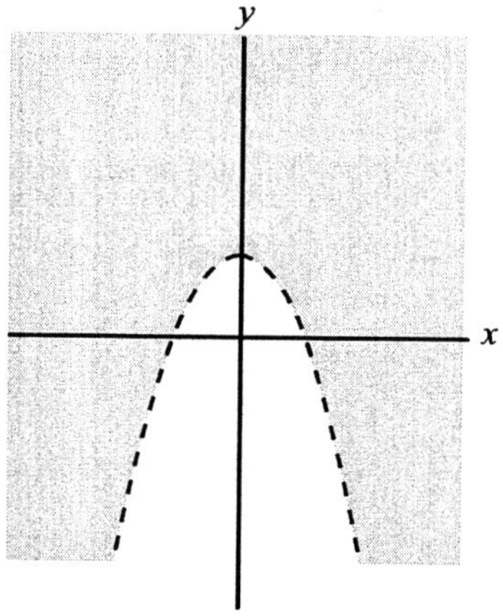

$y < 1 - x^2$

18.

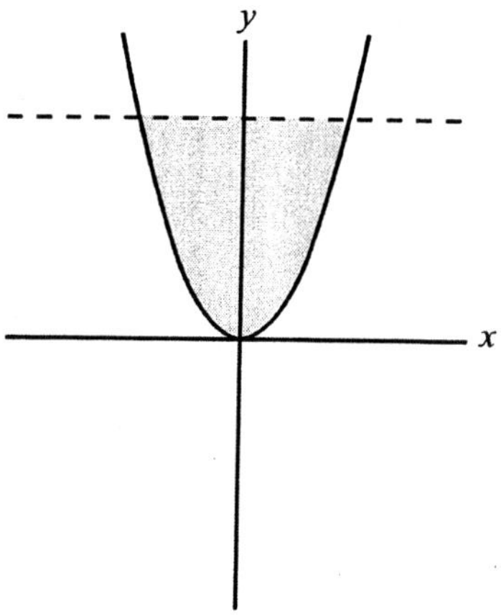

$y \geq x^2$
$y < 4$

20.

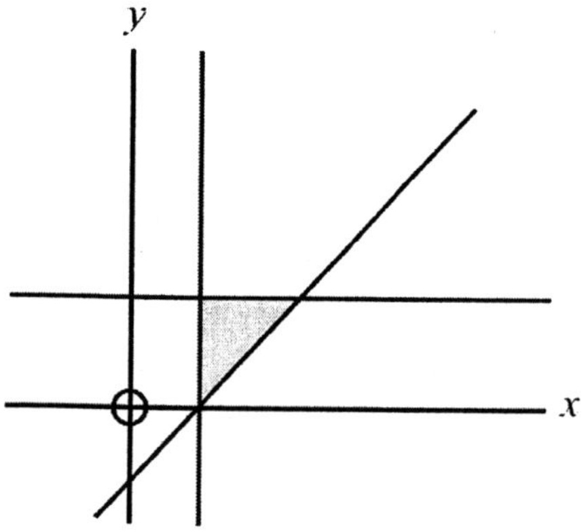

$y \geq x - 1$
$x \geq 1$
$y \leq 2$

22.

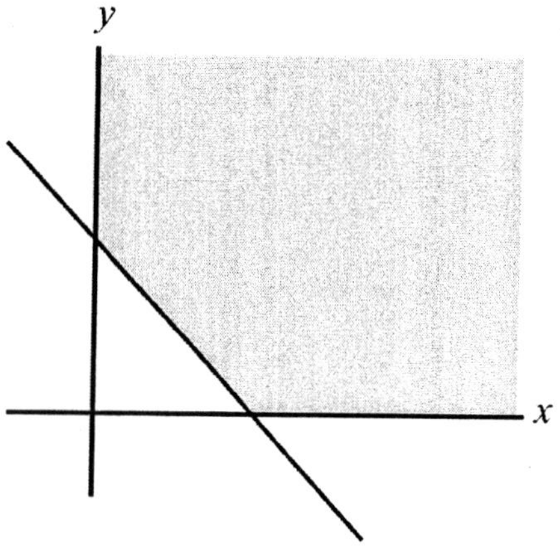

$y \geq 3 - x$
$x \geq 0$
$y \geq 0$

CHAPTER 18

Section 18.1

2. -27 4. $13\frac{2}{5}$ 6. 330 8. -105 10. $45\frac{3}{5}$ 12. 1530

14. $-4, -2, 0, 2, 4$ 16. $3, 2\frac{1}{2}, 2, 1\frac{1}{2}, 1$ 18. 9/7 20. 250, 500 22. \$8800; \$1600

Section 18.2

2. 8 4. 3/64 6. 768 8. 43/8 10. 765/64 or 11.953 12. 1023

14. $-6, -2, -2/3, -2/9, -2/27$ 16. $2, -3, 9/2, -27/4, 81/8$

18. $-5, 10, -20, 40, -80$ 20. 9 22. \$9,875 24. 15/16

26. 94.9 gal 28. 33/5 30. 9/10 32. -10 34. no limit

36. 5/37 38. 619/990 40. 365/999

Section 18.3

2. $16x^4 - 96x^3y + 216x^2y^2 - 216xy^3 + 81y^4$

4. $625x^4 + 500x^3 + 150x^2 + 20x + 1$

6. $1 + 7x + 21x^2 + 35x^3 + 35x^4 + 21x^5 + 7x^6 + x^7$

8. $a^8 - 16a^7b + 112a^6b^2 - 448a^5b^3 + 1120a^4b^4 - 1792a^3b^5 + 1792a^2b^6 - 1024ab^7 + 256b^8$

10. $\frac{243}{1024}m^5 + \frac{135}{128}m^4k + \frac{15}{8}m^3k^2 + \frac{5}{3}m^2k^3 + \frac{20}{27}mk^4 + \frac{32}{243}k^5$

12. $x^2 - 4x^{3/2}y^{1/2} + 6xy - 4x^{1/2}y^{3/2} + y^2$

14. $\frac{64x^6}{y^6} + \frac{576x^5}{y^5z} + \frac{2160x^4}{y^4z^2} + \frac{4320x^3}{y^3z^3} + \frac{4860x^2}{y^2z^4} + \frac{2916x}{yz^5} + \frac{729}{z^6}$

16. $2560x^3y^2$ 18. $15{,}360x^{3/2}$ 20. $70x^8y^{12}$ 22. $-8064x^5y^5$ 24. $56x^6$

CHAPTER 19

Section 19.1

2.

Bolts	0-4	5-9	10-14	15-19	20-24	25-29
Freq.	6	6	3	7	4	9

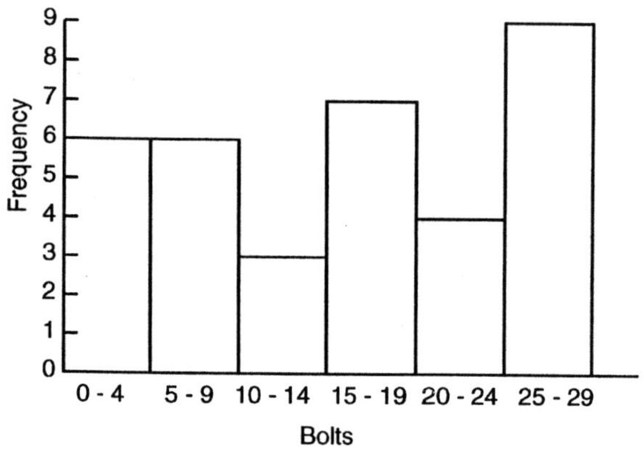

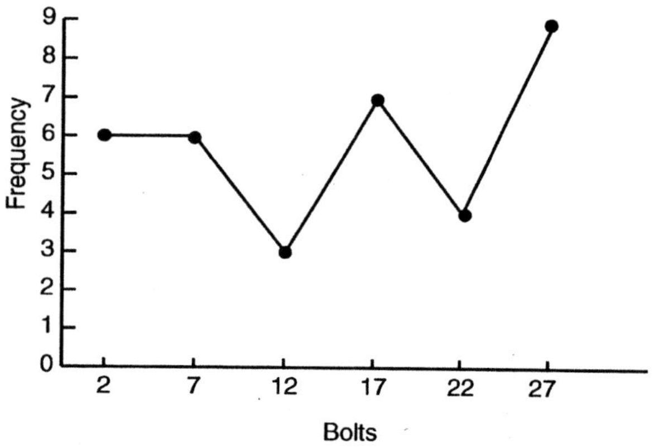

4.

Cheese	1-6	7-12	13-18	19-24	25-30	31-36	37-42	43-48
Freq.	2	7	1	7	3	3	0	2

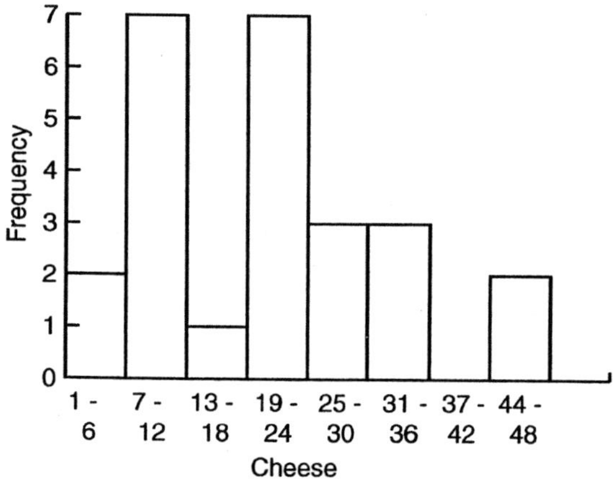

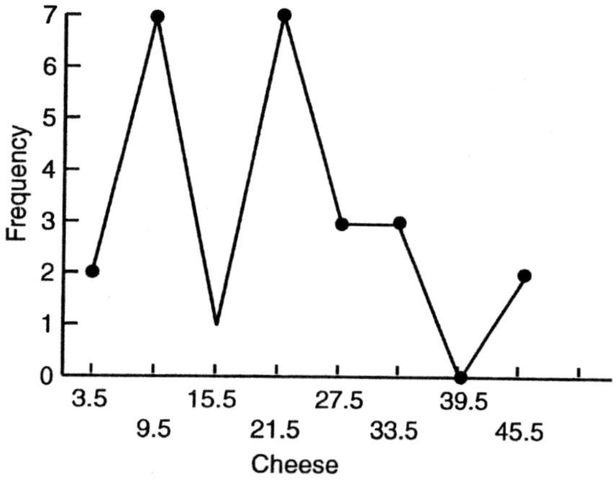

6.

Hamburgers	220-225	226-231	232-237	238-243	244-249	250-255	256-261	262-267	268-273	274-279
Freq.	3	3	1	2	2	5	2	0	5	2

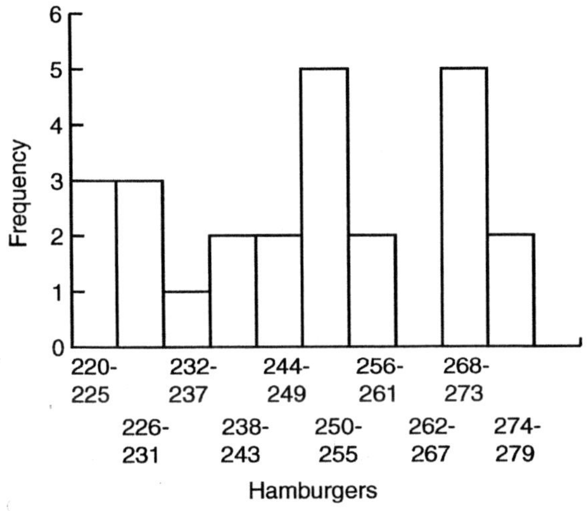

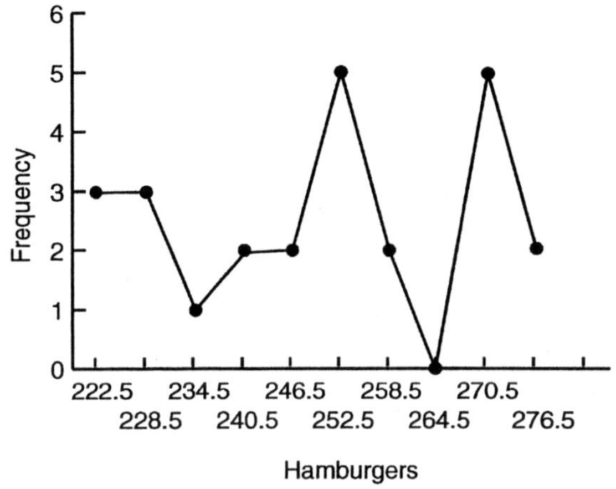

8.

Hours	205-219	220-234	235-249	250-264	265-279	280-294	295-309	310-324	325-339	340-354	355-369	370-384	385-399
Freq.	1	0	0	1	0	2	2	2	4	4	3	4	2

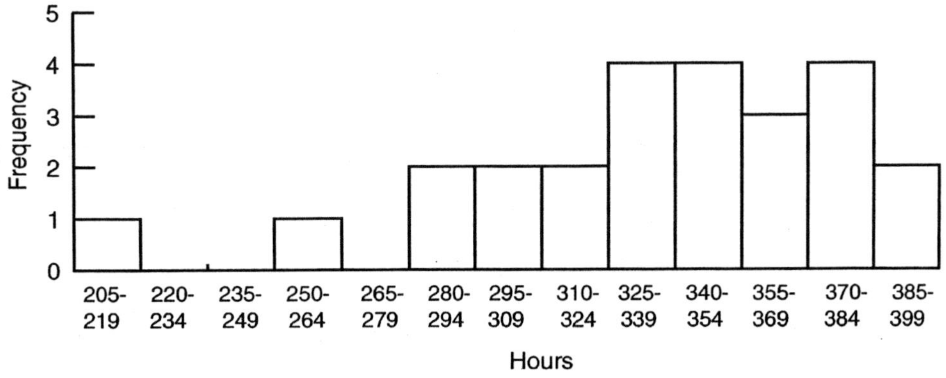

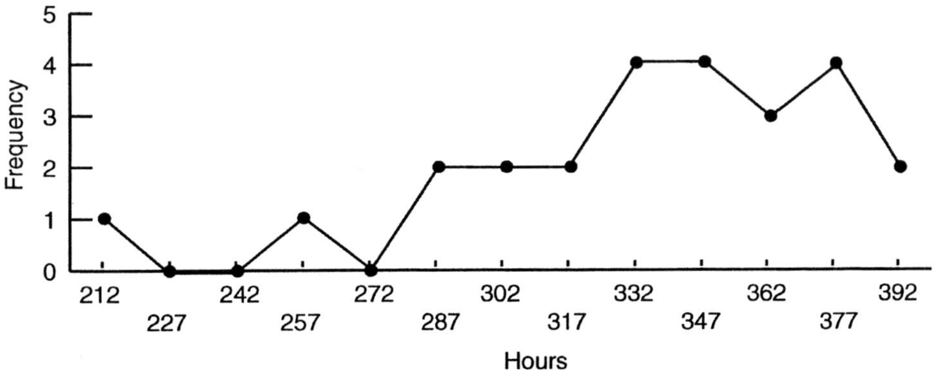

10.

Bars	1-4	5-8	9-12	13-16	17-20	21-24
Freq.	5	4	5	5	5	6

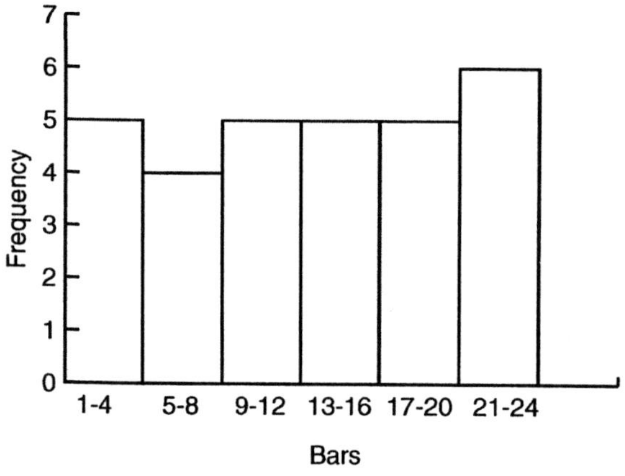

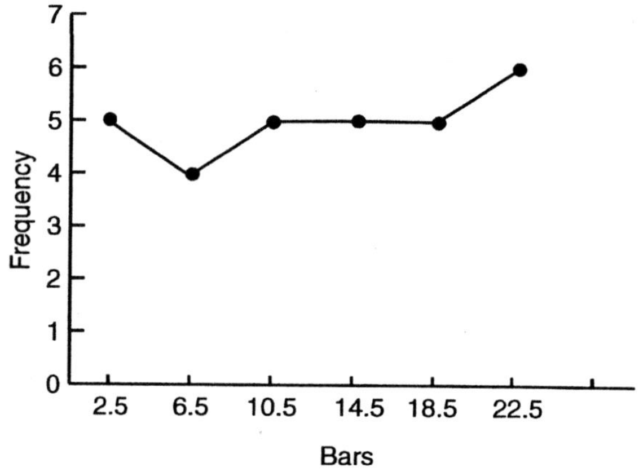

12.

Bags	0-1	2-3	4-5	6-7	8-9	10-11	12-13	14-15	16-17	18-19
Freq.	2	2	2	1	2	4	1	2	2	2

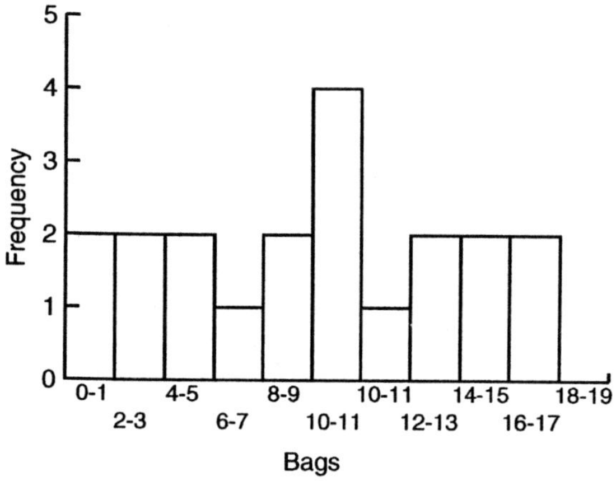

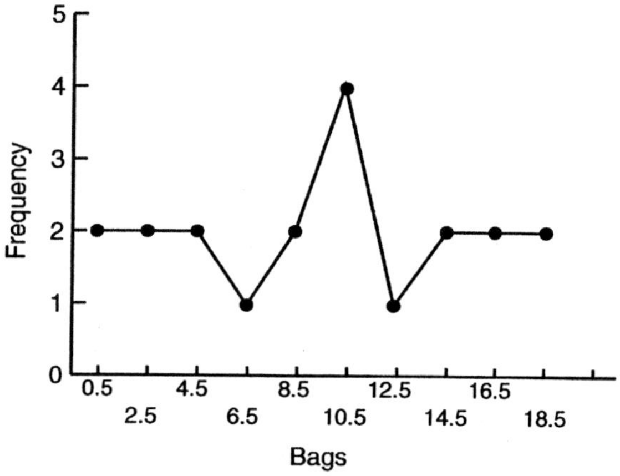

14.

Bottles	0-1	2-3	4-5	6-7	8-9	10-11	12-13	14-15	16-17
Frcq.	1	3	2	6	2	2	3	4	2

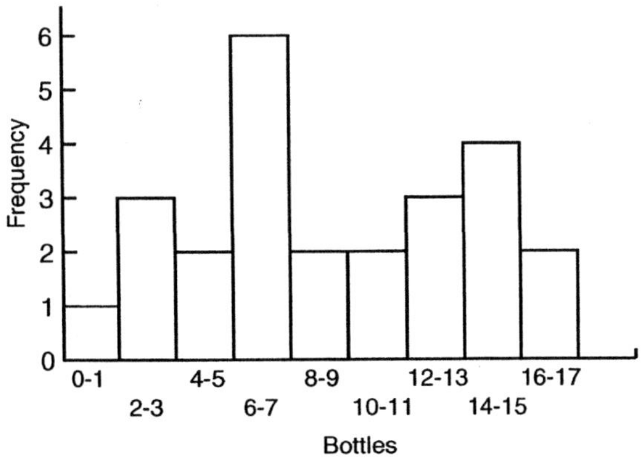

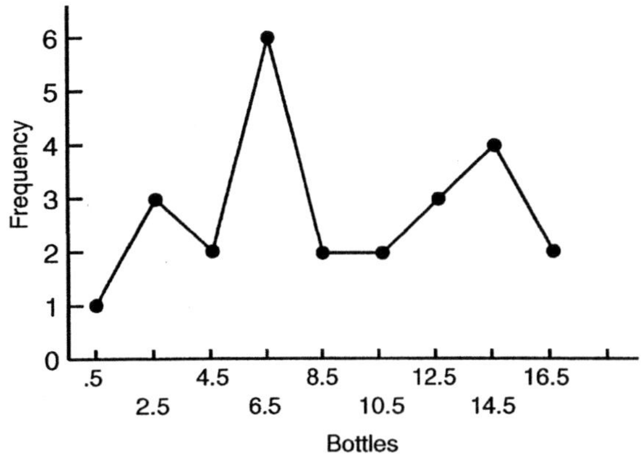

Section 19.2

2. 3.1, 2.85, 2.8 4. 8.98, 9.0, 9.0 6. 5.9, 6, 9

8. 344, none 10. 13, 21 12. 10, 16

14. 8, 7 16. 21 18. 9 20. 261, 268, 291

Section 19.3

2. 9, 3.23 4. 17, 6.35 6. 18, 6.39 8. 58, 6.30

10. 18, 6.98 12. 48.0, 16.1 14. 76.0, 33.6 16. 6.00, 2.01

18. 6.00, 1.78 20. 26.0, 8.22

Section 19.4

2. 53.1, 14.5, 75.0%, no 4. 61.4, 10.2, 66.7%, yes

6. 54.0, 9.08, 66.7%, yes 8. 66.7%, yes, 95.5%, yes

10. 68.0%, yes, 94.0%, yes 12. 71.0%, no, 98.1%, no 14. 2.5%

Section 19.5

2. $y = 1.4x + 3.7$ 4. $y = 0.34x + 0.78$ 6. $y = 1.1x^2 + 2.4x + 2.1$

8. $-0.93x^2 + 6.7x + 7.8$ 10. $y = -0.88t^2 + 8.0t + 11$

12. $y = 48300x^{-0.998}$ 14. $y = 215(0.223)^x$ or $215(4.48)^{-x}$

16. $y = 0.393x - 16.8$

Section 19.6

2. yes 4. no 6. yes 8. special cause

10. in control 12. ISO 9000

CHAPTER 20

Section 20.1

2.

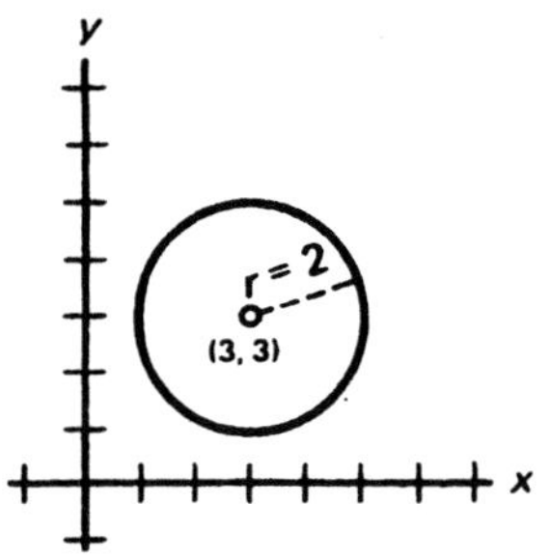

4.

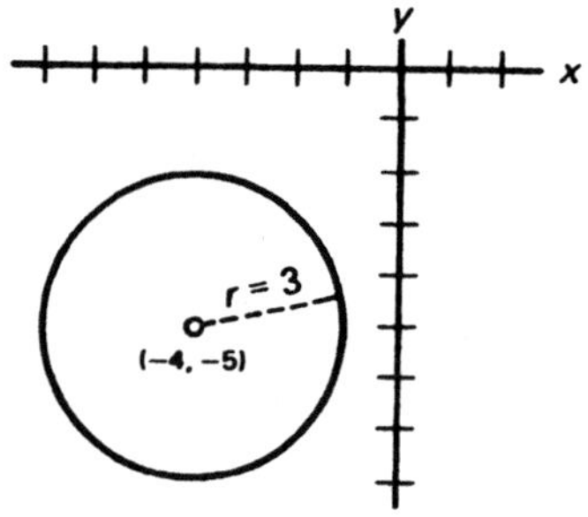

6. $(x + 2)^2 + (y - 3)^2 = 5$

8. $(x - 5)^2 + (y - 2)^2 = 113$

10. $x^2 + y^2 = 25$

12. $(2, 0)$; $r = 3$

14. $(3, -7)$; $r = 4$

16. $(-5, -1)$; $r = 2\sqrt{10}$

18. $(-2, 9/2)$; $r = 9/2$

20. $(5/2, 4)$; $r = \sqrt{89}/2$

22. $x^2 + y^2 - 6x - 12y + 9 = 0$; $(3, 6)$

24. $x^2 + y^2 - 2x - 2y - 23 = 0$; $(1, 1)$, $r = 5$

Section 20.2

2. $x^2 = -8y$

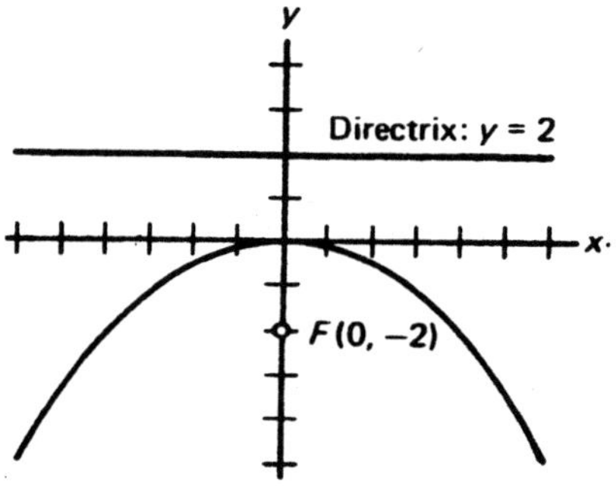

4. $x^2 = -6y$

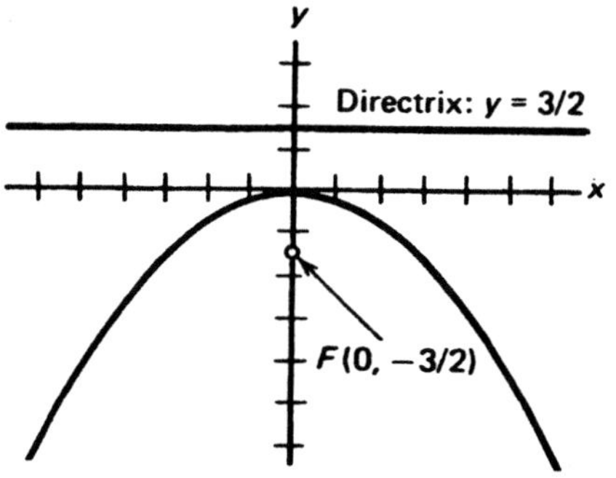

6. $y^2 = -4x$

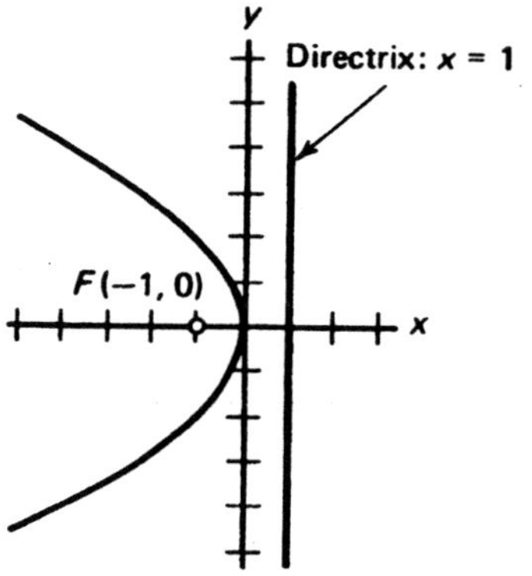

8. $y^2 = -12x$

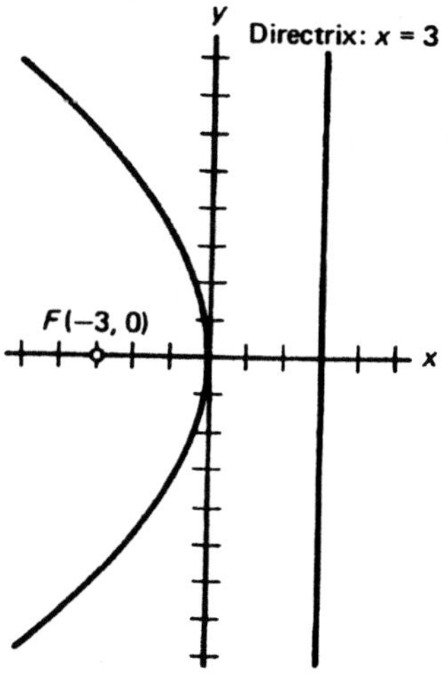

10. $x^2 = -y$

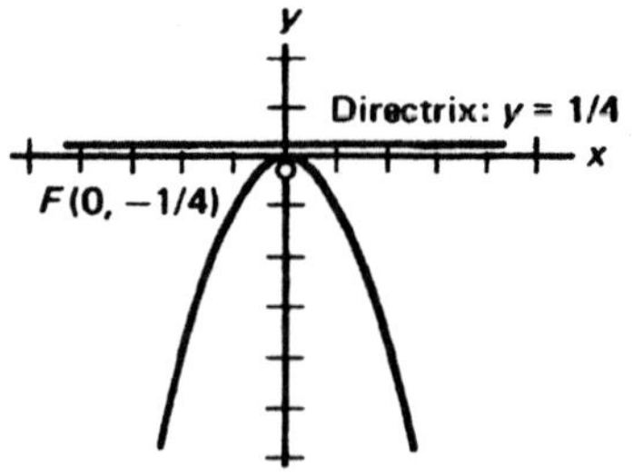

12. $x^2 = -12y$

14. $y^2 = 20x$

16. $x^2 = -4y$

18. $x^2 = 8y$

20. $x^2 - 4x + 8y + 28 = 0$

22. $x^2 = 10y$

24. 73.5 cm

26. $V(-3, 1), x = -3$

28. $V(-1, 7), x = -1$

30. a. 22,500 ft
    b. 37.5 s

32. 9 in. by 18 in.

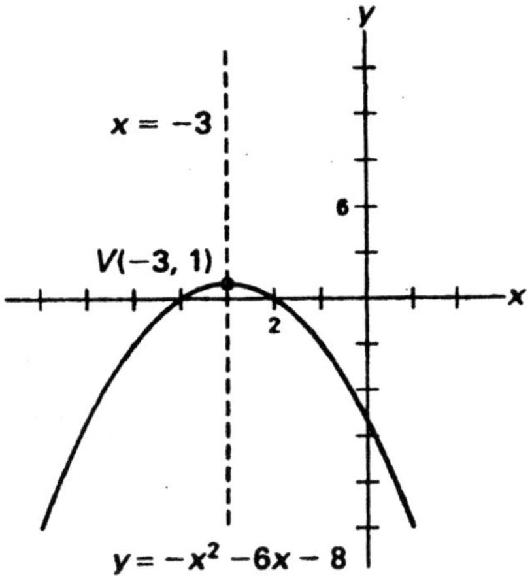

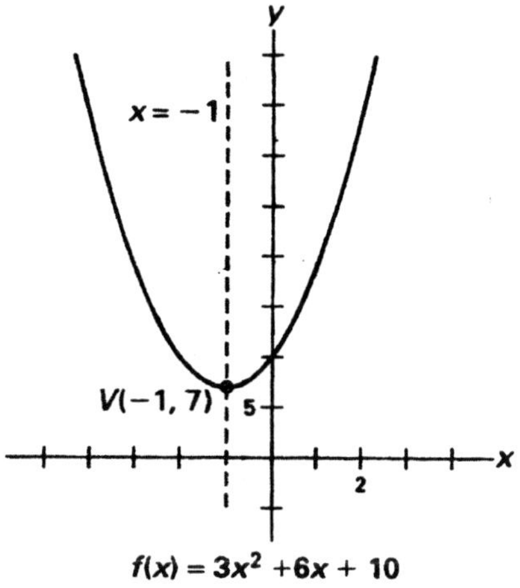

Section 20.3

2. $V(0, \pm 8)$; $F(0, \pm 2\sqrt{7})$;
   16, 12; $x^2/36 + y^2/64 = 1$

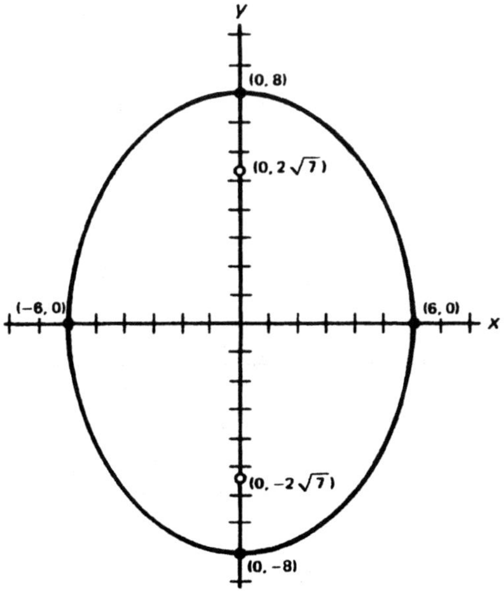

4. $V(0, \pm 5)$; $F(0, \pm 3)$; 10, 8
   $25x^2 + 16y^2 = 400$

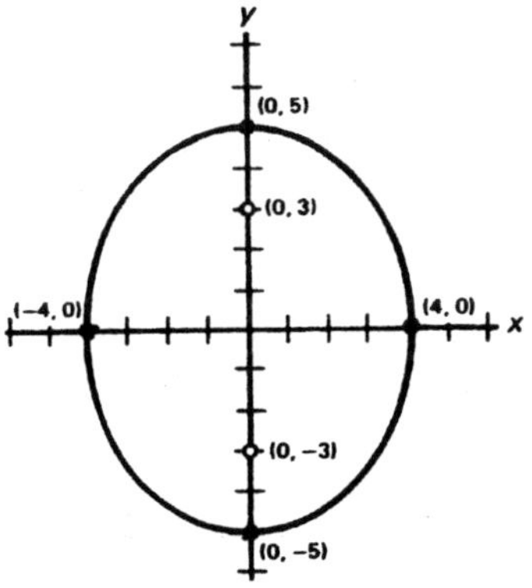

6. $V(0, \pm 2)$; $F(0, \pm 1)$; 4, $2\sqrt{3}$
   $4x^2 + 3y^2 = 12$

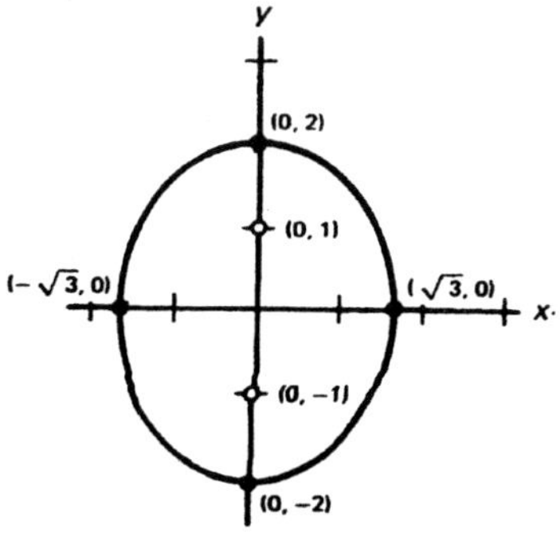

8. $V(\pm 4, 0)$; $F(\pm 2\sqrt{3}, 0)$; 8, 4
   $x^2 + 4y^2 = 16$

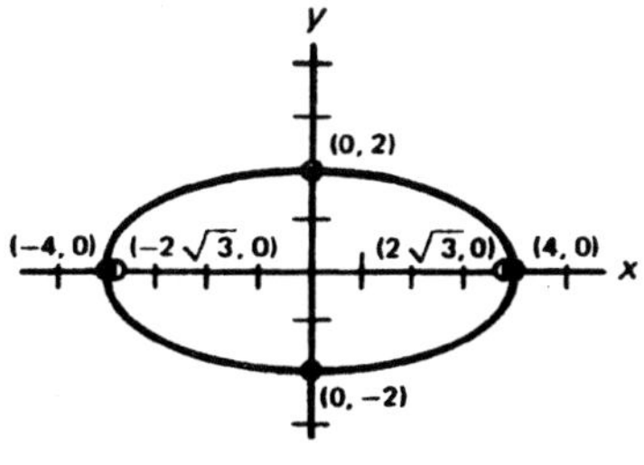

10. $x^2/24 + y^2/49 = 1$ or $49x^2 + 24y^2 = 1176$

12. $x^2/144 + y^2/44 = 1$ or $11x^2 + 36y^2 = 1584$

14. $x^2/81 + y^2/100 = 1$ or $100x^2 + 81y^2 = 8100$

16. $x^2/16 + y^2/7 = 1$ or $7x^2 + 16y^2 = 112$

18. $x^2/400 + y^2/64 = 1$ or $4x^2 + 25y^2 = 1600$

Section 20.4

2. $V(\pm 12, 0)$; $F(\pm 13, 0)$;
   24, 10; $y = \pm 5x/12$
   $x^2/144 - y^2/25 = 1$

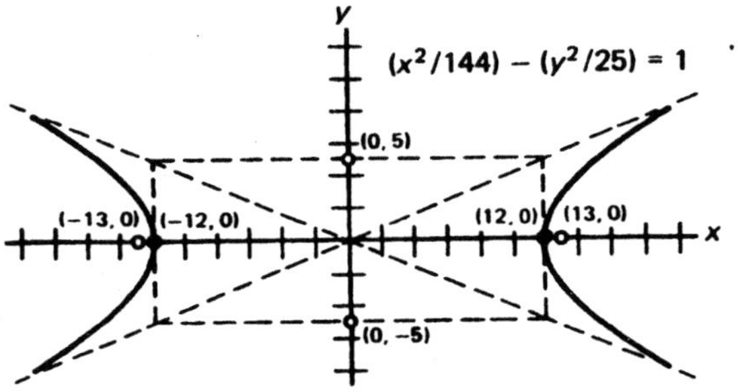

4. $V(0, \pm 4)$; $F(0, \pm 5)$
   8, 6; $y = \pm 4x/3$
   $y^2/16 - x^2/9 = 1$

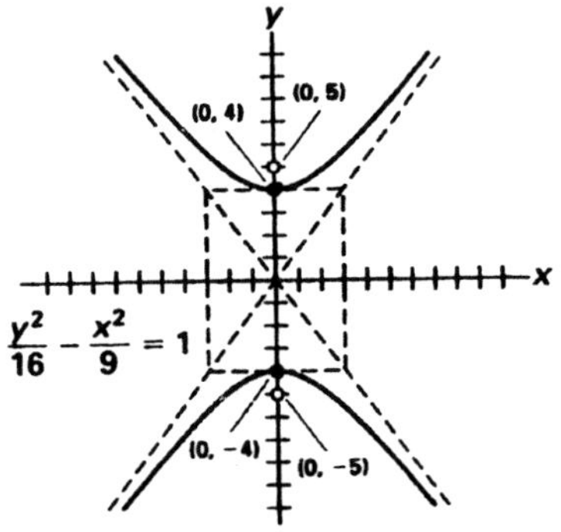

6. $V(0, \pm\sqrt{2})$; $F(0, \pm\sqrt{5})$
   $2\sqrt{2}, 2\sqrt{3}$;
   $y = \pm\sqrt{(2/3)}x$

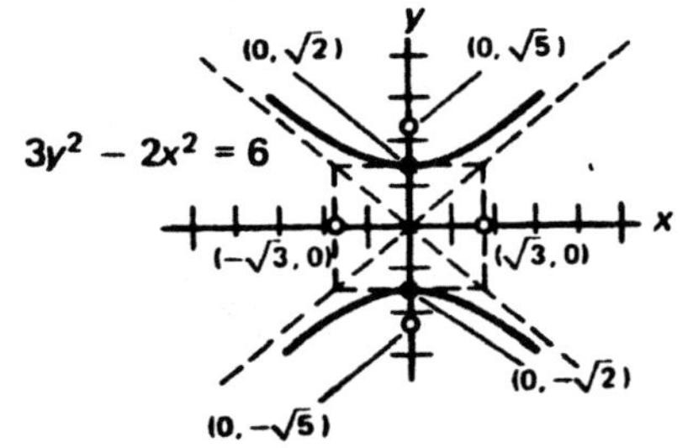

8. $V(\pm 1, 0)$; $F(\pm\sqrt{5}, 0)$
   2, 4; $y = \pm 2x$

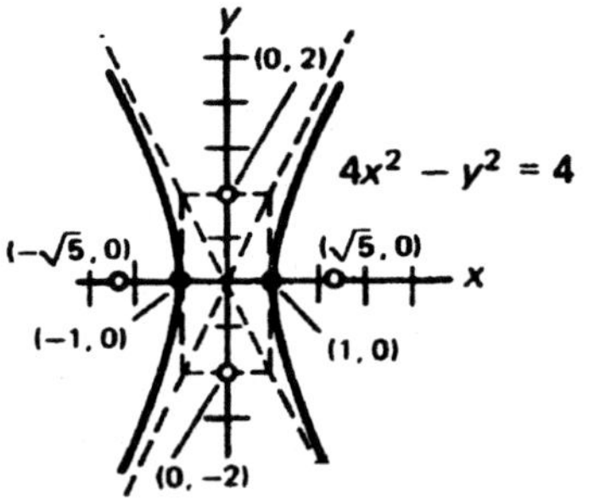

10. $y^2/25 - x^2/24 = 1$ or $24y^2 - 25y^2 = 600$

12. $x^2/4 - y^2/21 = 1$ or $21x^2 - 4y^2 = 84$

14. $y^2/36 - x^2/16 = 1$ or $4y^2 - 9x^2 = 144$

16. $y^2/36 - x^2/28 = 1$ or $7y^2 - 9x^2 = 252$

18. $xy = -4$

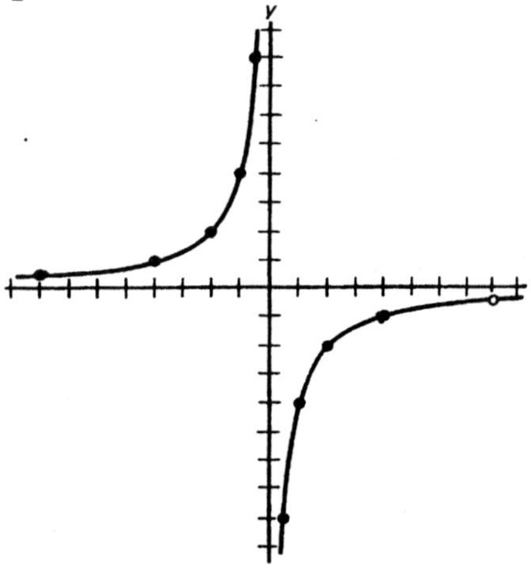

Section 20.5

2. $(x+1)^2 = 4(y-3)$

4. $\dfrac{(x+2)^2}{36} + \dfrac{(y+3)^2}{25} = 1$

6. $\dfrac{(x+2)^2}{9} - \dfrac{(y+2)^2}{25} = 1$

8. ellipse; $C(-1, 2)$

10. parabola; $V(-1, 0)$

12. hyperbola; $C(-1, -3)$

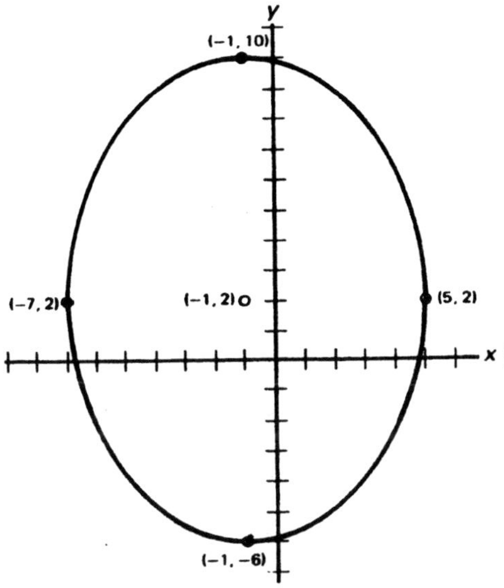

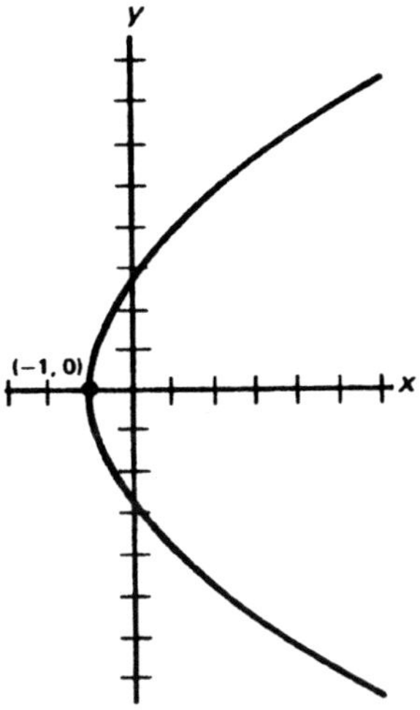

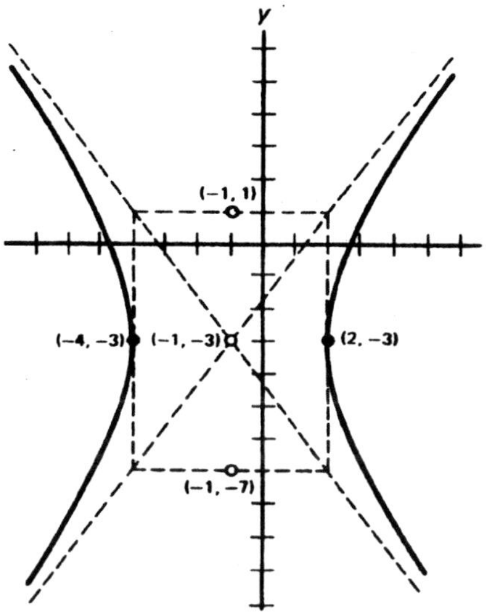

14. hyperbola; $C(3, 1)$

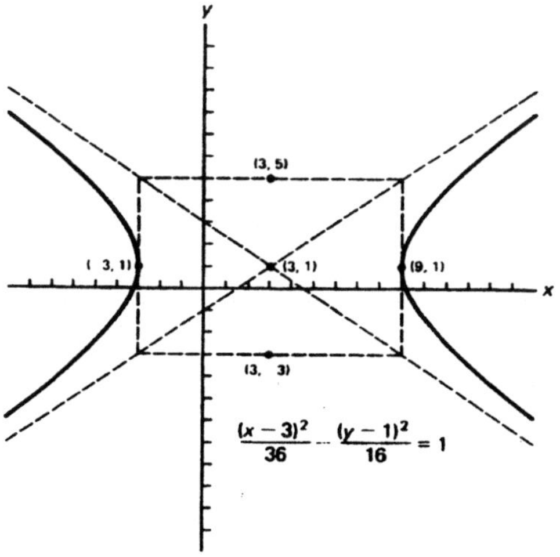

16. parabola; $V(5, -2)$

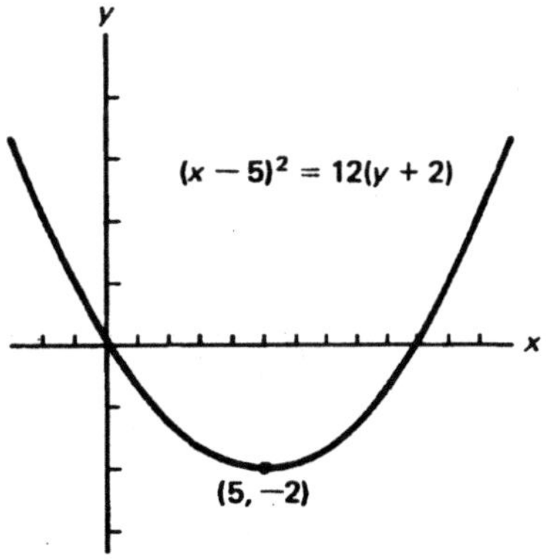

18. ellipse; $C(2, -4)$

20. ellipse; $C(1, -3)$

22. hyperbola; $C(2, 0)$

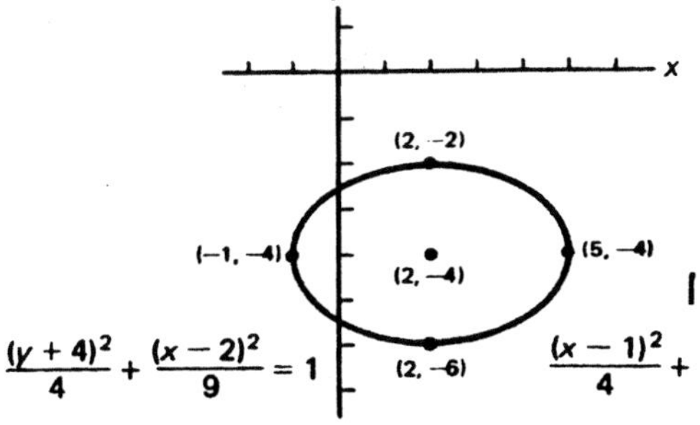

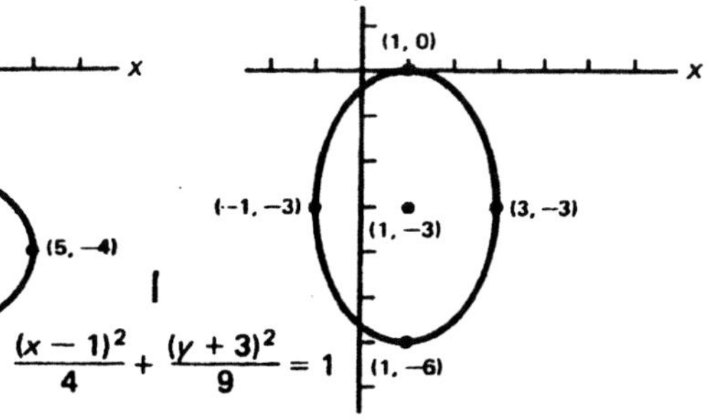

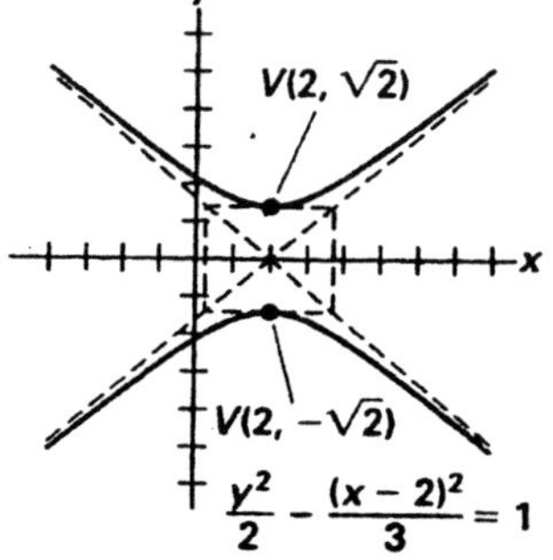

24. parabola; $V(3, -3)$

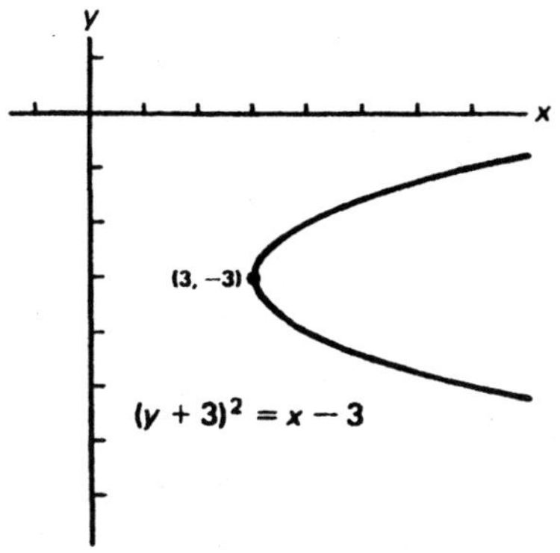

26. ellipse; $C(2, 3)$

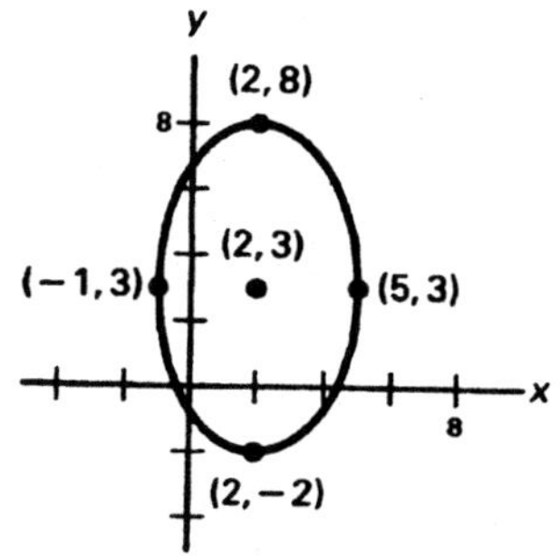

$$\frac{(x-2)^2}{9} + \frac{(y-3)^2}{25} = 1$$

28. hyperbola; $C(-3, 5)$

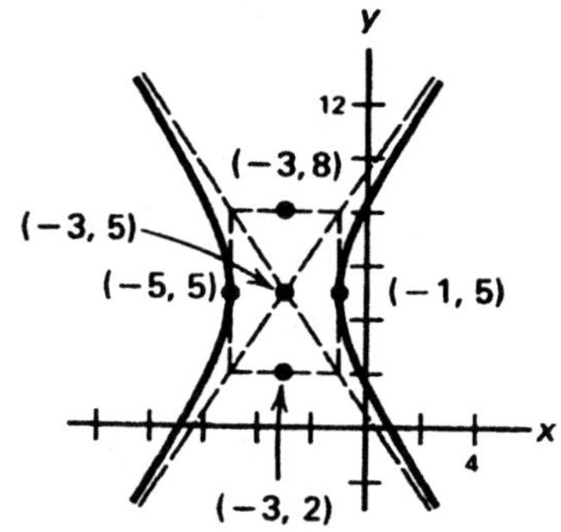

$$\frac{(x+3)^2}{4} + \frac{(y-5)^2}{9} = 1$$

30. parabola; $V(3, 3)$

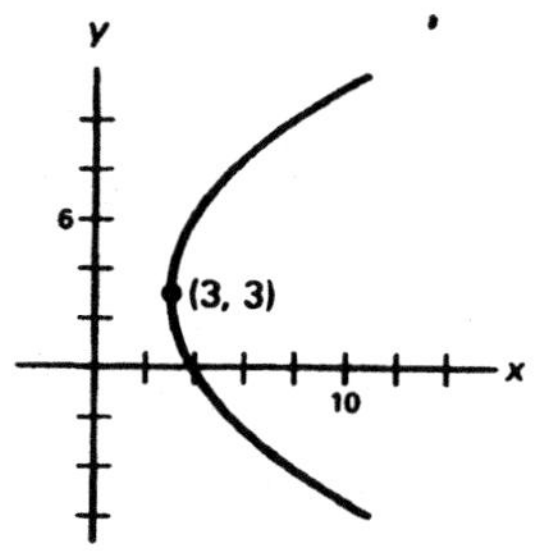

$$(y-3)^2 = 10(x-3)$$

Section 20.6

2. circle
4. ellipse
6. circle
8. parabola
10. hyperbola
12. hyperbola
14. circle
16. parabola

Section 20.7

2. $(1, 0)$ and $(-1, 0)$
4. $(2, -2)$ and $(-2, 2)$
6. no solution
8. $(0, 0)$ and $(3, 3)$
10. $(3, 0)$ and $(0, 4)$
12. $(2, \sqrt{2})$ and $(2, -\sqrt{2})$
14. $(-2, 0)$ and $(3, -10)$
16. $\left(\frac{2\sqrt{7}}{\sqrt{3}}, \frac{2\sqrt{5}}{\sqrt{3}}\right), \left(\frac{2\sqrt{7}}{\sqrt{3}}, \frac{-2\sqrt{5}}{\sqrt{3}}\right), \left(\frac{-2\sqrt{7}}{\sqrt{3}}, \frac{2\sqrt{5}}{\sqrt{3}}\right), \left(\frac{-2\sqrt{7}}{\sqrt{3}}, \frac{-2\sqrt{5}}{\sqrt{3}}\right)$

Section 20.8

2.

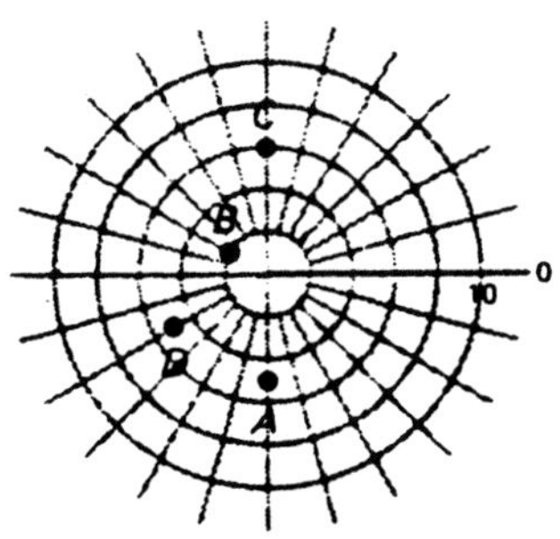

4.

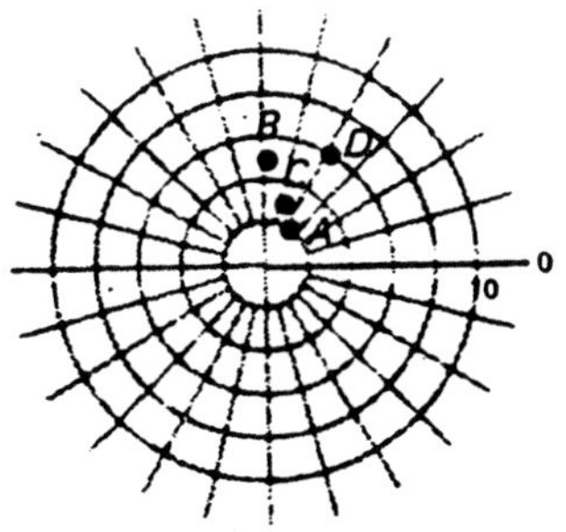

6. (–2, 60°), (2, –120°), (–2, –300°)

8. (6, 270°), (6, –90°), (–6, –270°)

10. (1, 0°), (–1, 180°), (1, 360°)

12. (7, $3\pi/2$), (7, $-\pi/2$), (–7, $-3\pi/2$)

14. (2, $-11\pi/6$), (2, $\pi/6$), (–2, $7\pi/6$)

16. (–5, $-2\pi/3$), (5, $\pi/3$), (–5, $4\pi/3$)

18.

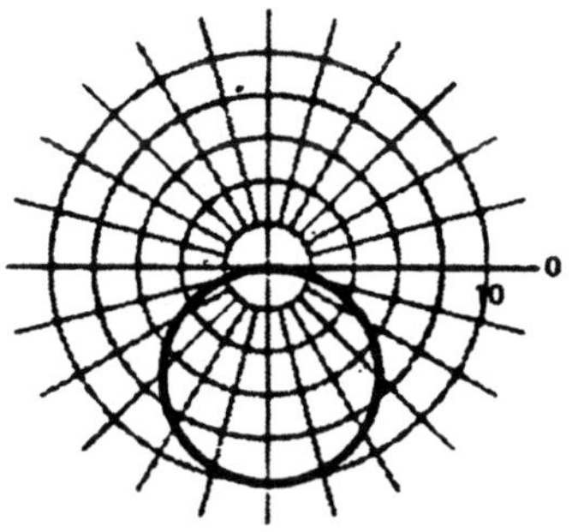

$r = -10 \sin \theta$

20.

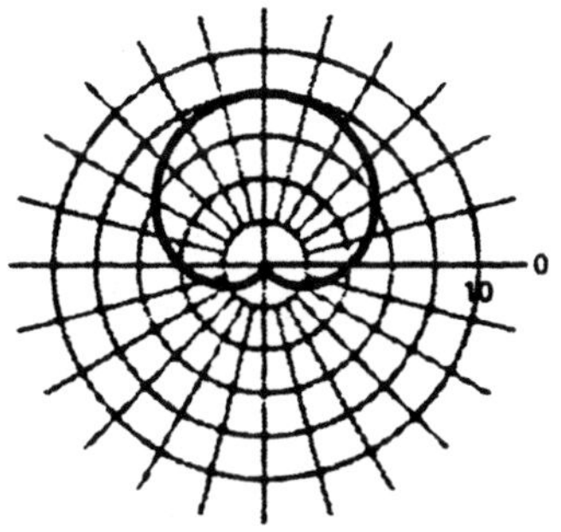

$r = 4 + 4 \sin \theta$

22.

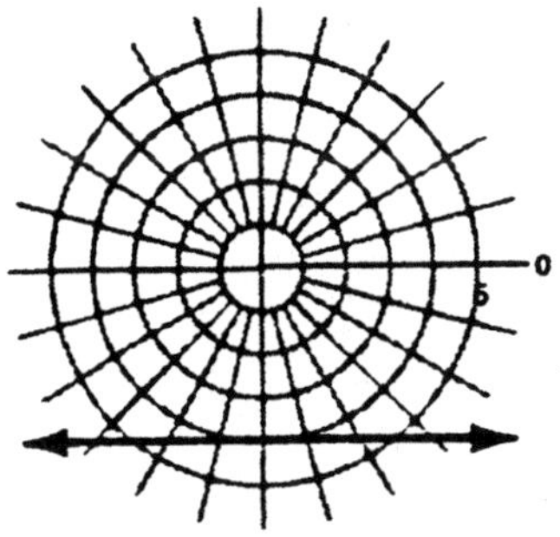

$r \sin \theta = -4$

24.

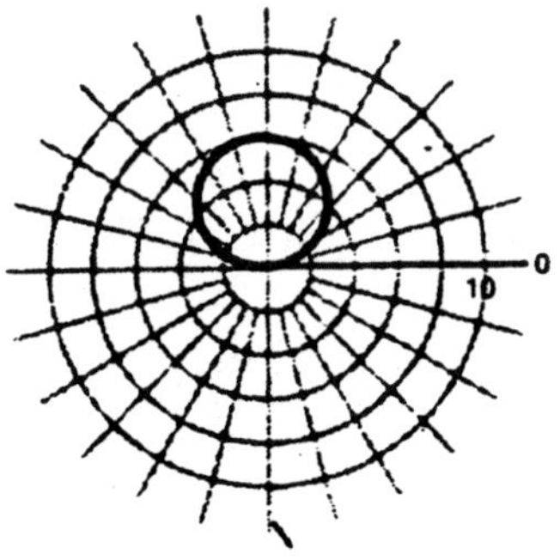

$r = 6 \sin \theta$

26.

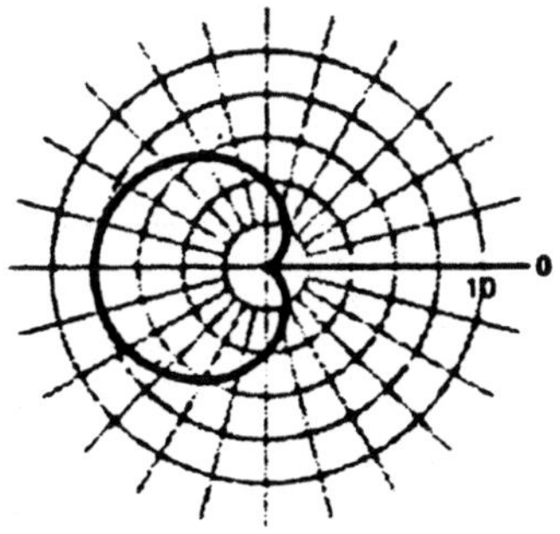

$r = 4 - 4 \cos \theta$

28.

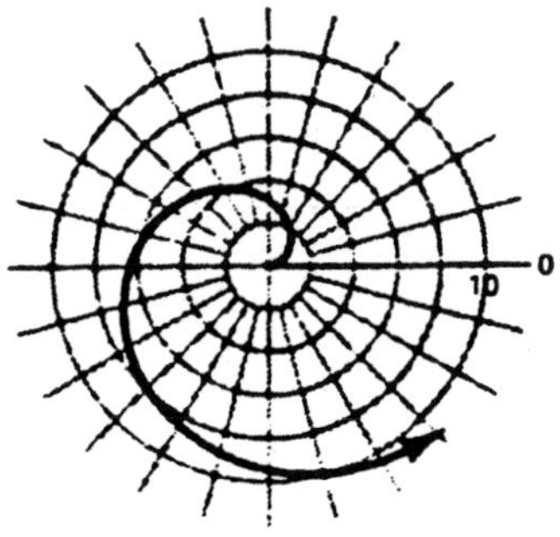

$r = 2\theta$

30. (–2, 0)

32. $(-7\sqrt{3}/2, 7/2)$

34. $(1/2, \sqrt{3}/2)$

36. (–3, 0)

38. $(-1, -\sqrt{3})$

40. $(-1/2, -\sqrt{3}/2)$

42. (2, 150°)

44. (3, 180°)

46. ($\sqrt{2}$, 135°)

48. (2, $4\pi/3$)

50. (10, $7\pi/4$)

52. (0, θ)

54. $r \sin \theta = 5$

56. $r \sin^2\theta = 5 \cos \theta$

58. $r = \dfrac{6}{2\cos\theta + 3\sin\theta}$

60. $r = \dfrac{6}{a\cos\theta + b\sin\theta}$

62. $r^2 = \dfrac{36}{4 - 13\sin^2\theta}$

64. $\tan^2\theta = 1$

66. $x = 7$

68. $x = 3$

70. $y = \sqrt{3}\,x$

72. $x^2 + y^2 - 6y = 0$

74. $x^2 + y^2 + 2\sqrt{2}\,x - 2\sqrt{2}\,y = 0$

76. $x^4 + x^2y^2 - y^2 = 0$

78. $x^2 - y^2 = 6$

80. $x^4 + 2x^2y^2 + y^4 - x^2 + y^2 = 0$

82. $x^2 = 4y$

84. $y^2 = 8(x + 2)$

86. $(x + y)^{3/2} = 4(x^2 - y^2)$

88. $x^4 + 2x^2y^2 + y^4 + 2x^3 + 2xy^2 - y^2 = 0$

90. $\sqrt{31}$

Section 20.9

2.

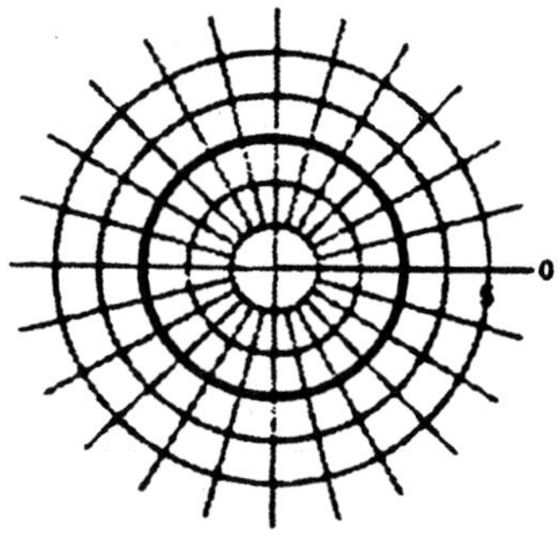

$r = 3$

4.

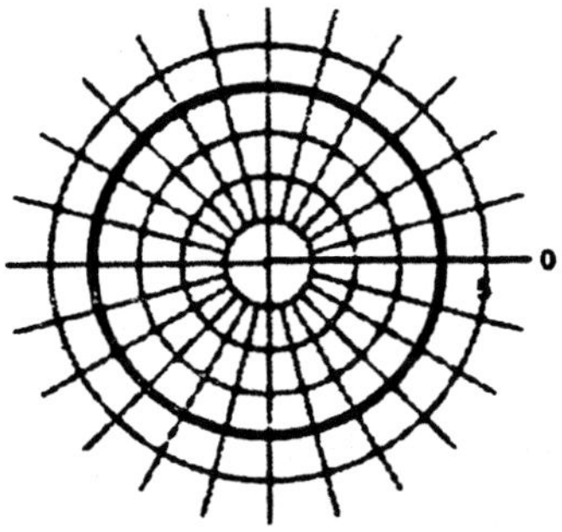

$r = -4$

6.

$\theta = -120°$

8.

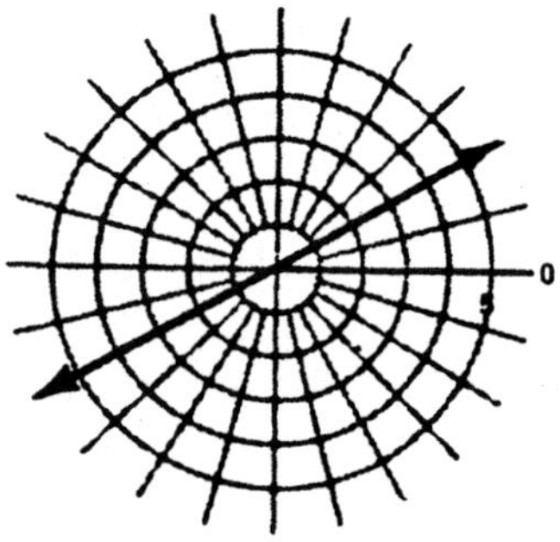

$\theta = 7\pi/6$

10.

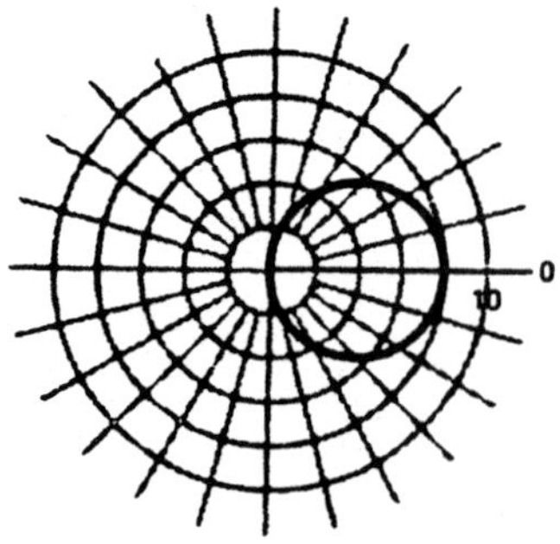

$r = 8\cos\theta$

12.

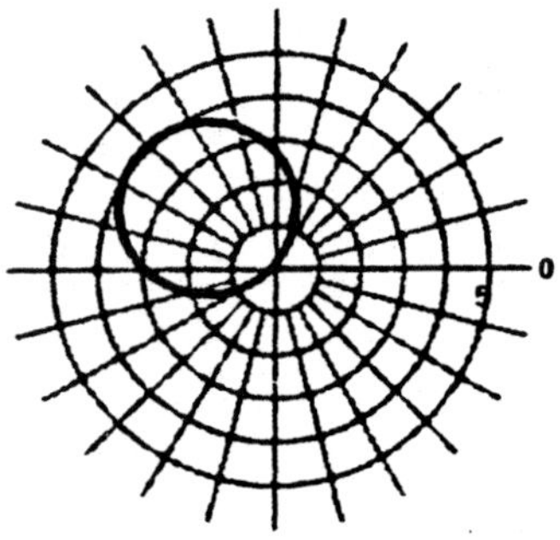

$r = 4\sin(\theta - \pi/4)$

14.

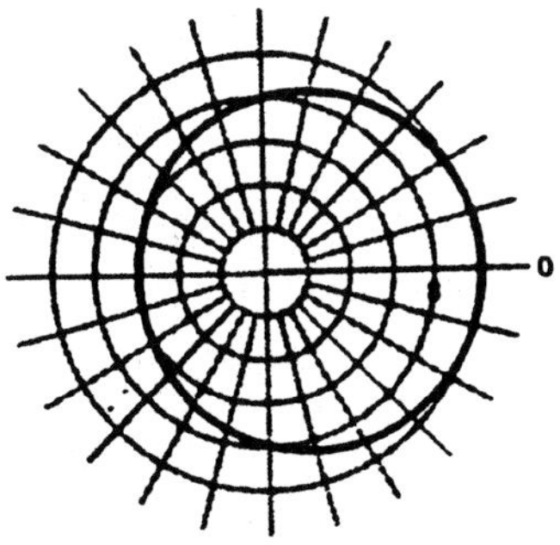

$r = 8 + 2\cos\theta$

16.

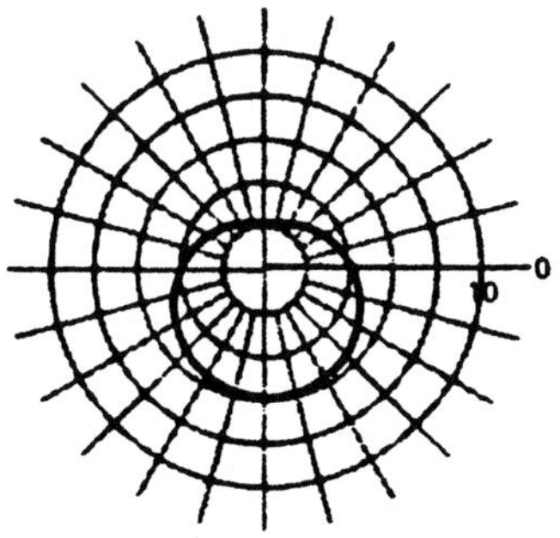

$r = 4 - 2\sin\theta$

18.

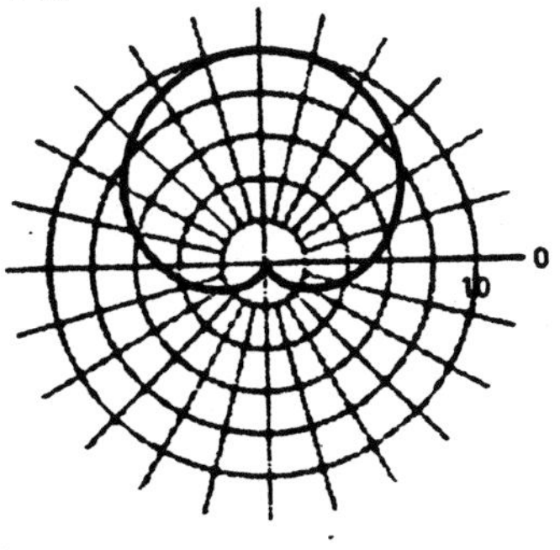

$r = 5 + 5\sin\theta$

20.

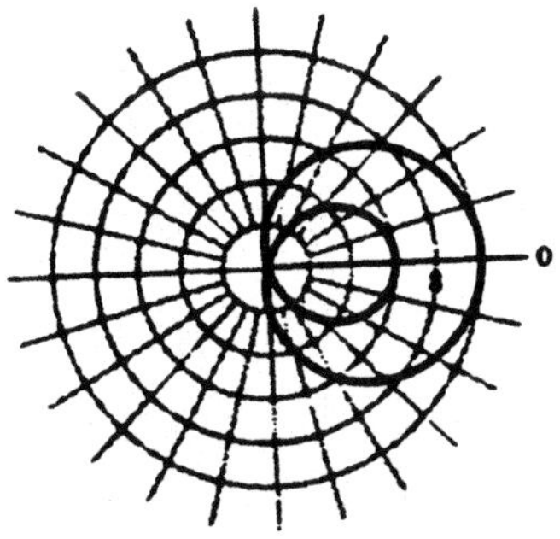

$r = 2 + 8\cos\theta$

22.

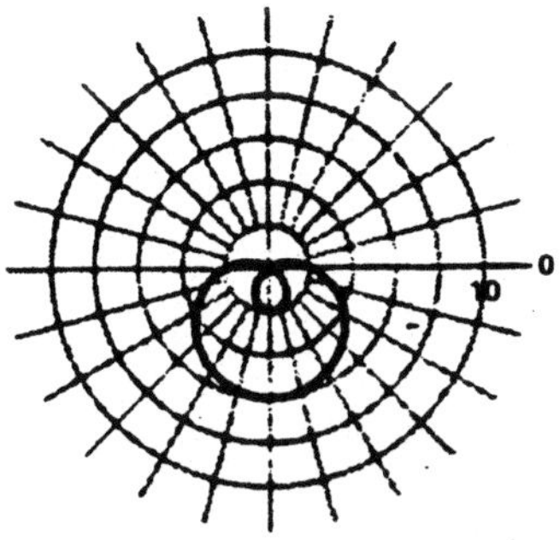

$r = 2 - 4\sin\theta$

24.

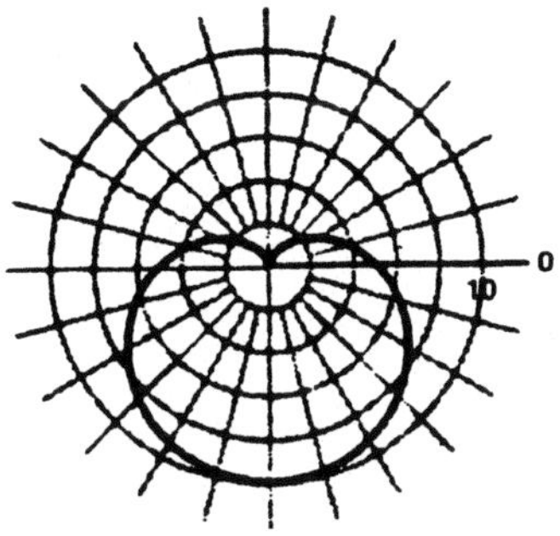

$r = 5 - 5\sin\theta$

26.

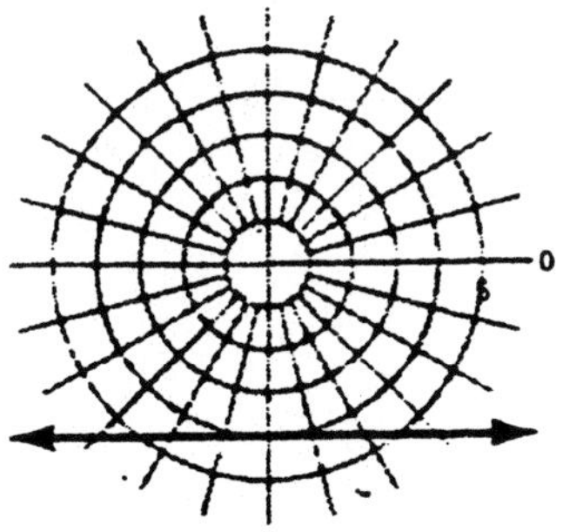

$r\sin\theta = -4$

28.

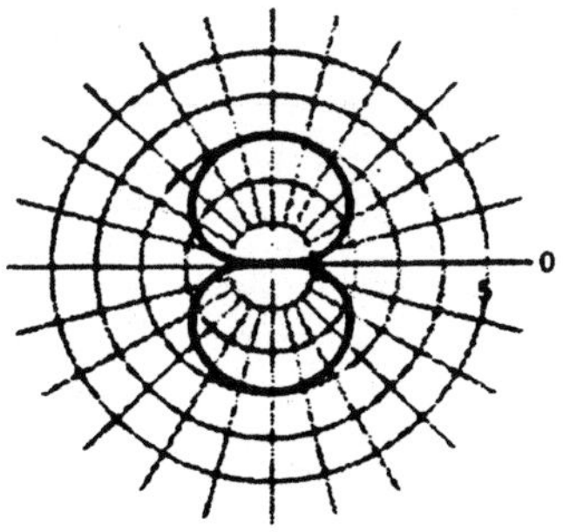

$r^2 = -9\sin\theta$

30.

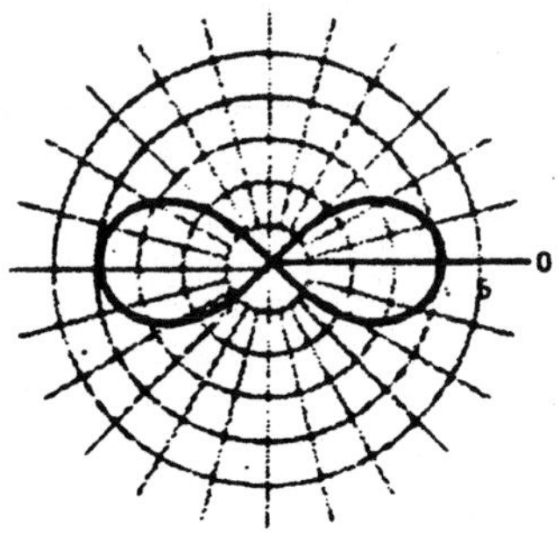

$r^2 = 16\cos 2\theta$

32.

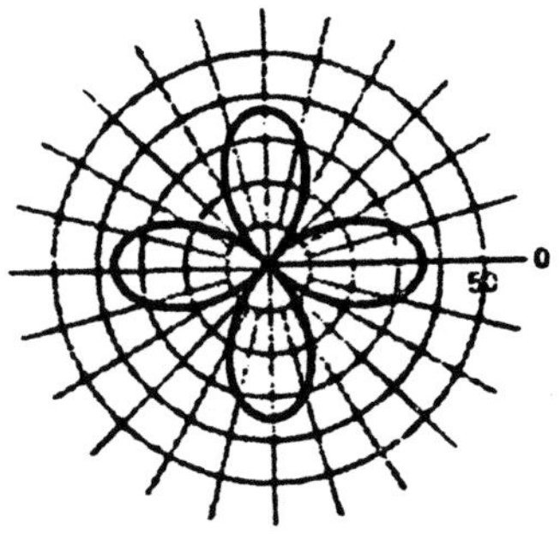

$r = -36 \cos 2\theta$

34.

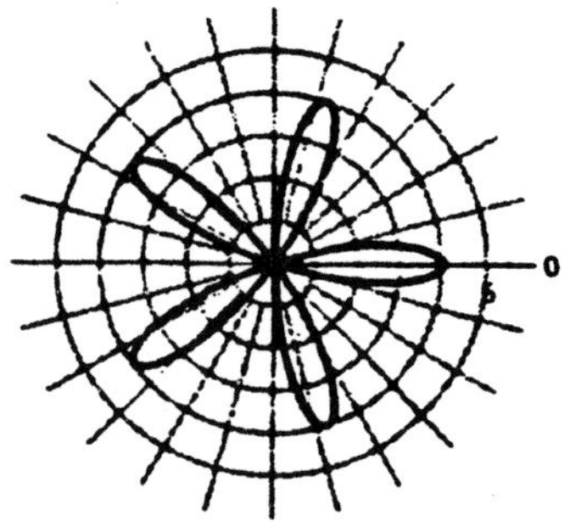

$r = 4 \cos 5\theta$

36.

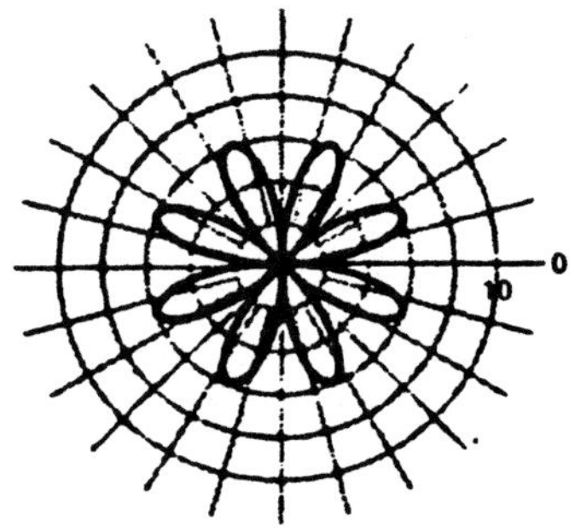

$r = 6 \sin 4\theta$

38.

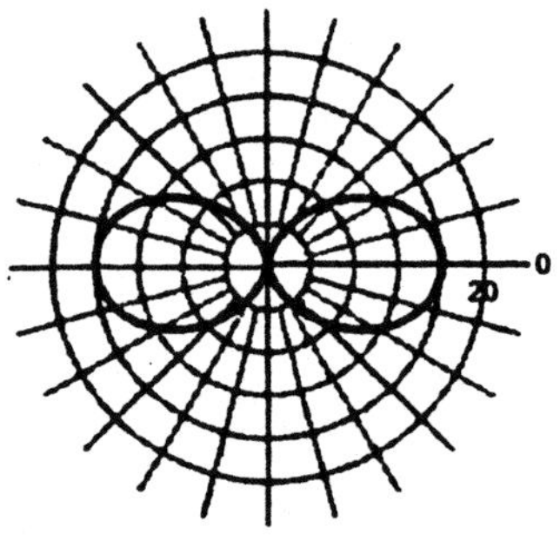

$r = 16 \cos^2 \theta$

40.

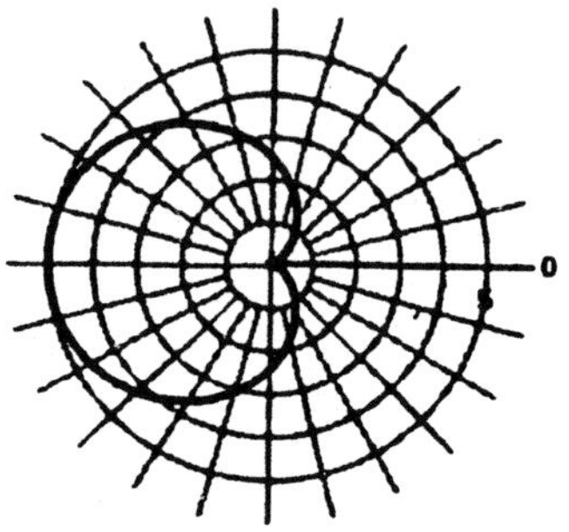

$r = 5 \sin^2 (\theta/2)$

42.

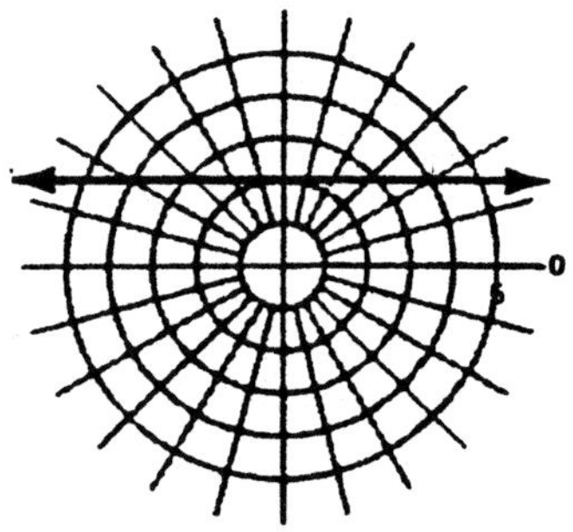

$r = 2 \csc \theta$

44.

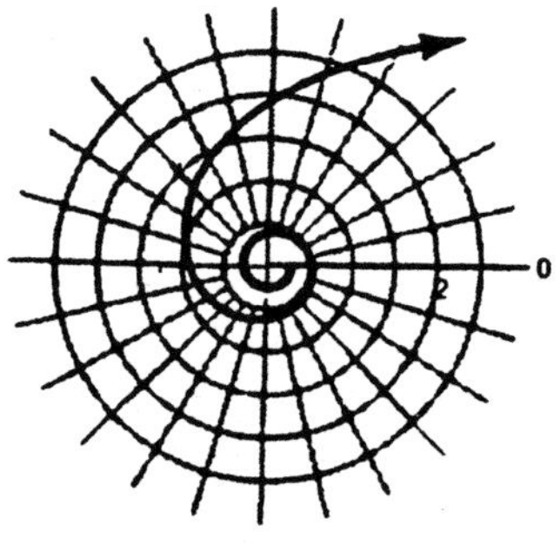

$r = \frac{3}{\theta},\ (\theta > 0)$

46.

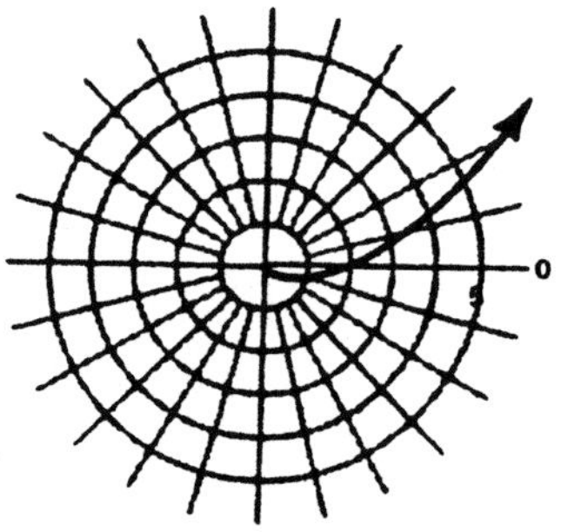

$r = 2 \cdot 3^{2\theta}$

48.

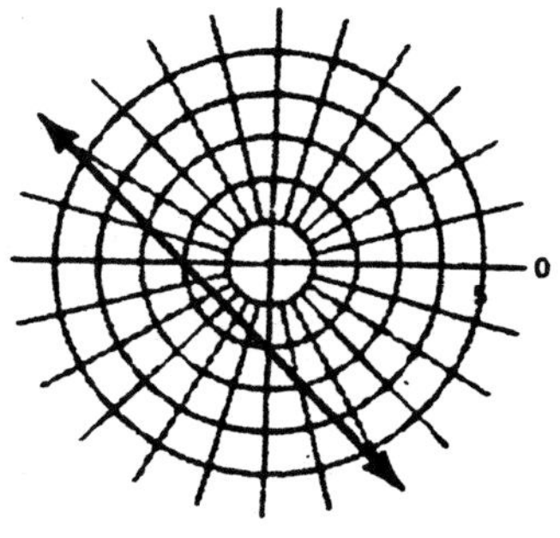

$r = \dfrac{-2}{\sin \theta + \cos \theta}$

50.

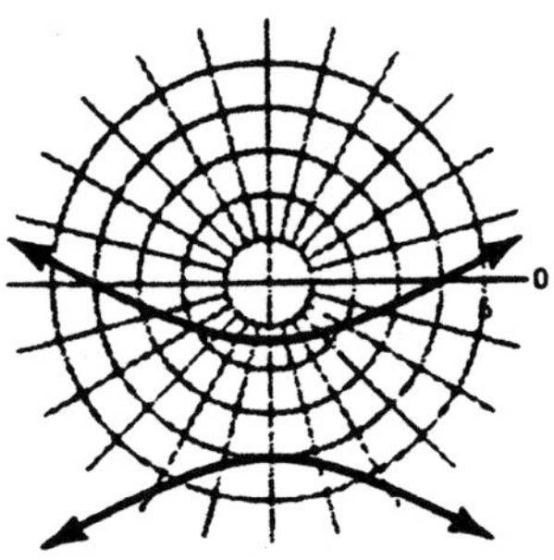

$r(1 + 2 \sin \theta) = -4$

CHAPTER 21

Section 21.1

2. 3   4. 13   6. −24   8. $\sqrt{15}$ 10. $w^2 + 9w + 17$

12. $(\Delta t)^2 - 9(\Delta t) + 22$ 14. $2(\Delta t)$; 2

16. $3(\Delta t)^2 + 6t(\Delta t)$; $3(\Delta t) + 6t$

18. $5(\Delta t)^2 + 10t(\Delta t) + 3(\Delta t)$; $5(\Delta t) + 10t + 3$ 20. 14 m/s

22. 68 m/s   24. 2×10^5 m/s   26. 240μA   28. −24 m/s   30. 67 m/s

32. −1/9 m/s   34. −1/48 m/s   36. 6,000 cm/s

Section 21.2

2. −2   4. 3/2   6. 0   8. −3   10. 2   12. −6

14. 8   16. 3   18. no limit   20. 0   22. 7/4   24. 13

26. 13   28. 36   30. −10   32. 9   34. 120   36. 2/15

38. 4   40. −8   42. 15   44. $3a^2$ 46. $-2/a^3$ 48. $1/(2\sqrt{x})$

50. b   52. b   54. b   56. b   58. yes   60. no

62. yes   64. yes   66. b   68. c   70. 0   72. c

Section 21.3

2. 6   4. 16   6. −4   8. −23   10. 3   12. $y = -20x + 20$

14. $y = 8x + 6$ 16. $y = 17x - 31$ 18. $y = 12x + 1$

20. $y = 9x - 12$ 22. (1, −1)   24. (3, −15), (−1, 1)

Section 21.4

2. −6   4. 4   6. $-8x$ 8. $2x$ 10. $12x - 8$ 12. $-4 - 14x$ 14. $-6x^2$

16. $\dfrac{-1}{(x-1)^2}$ 18. $\dfrac{2}{(2x+1)^2}$ 20. $\dfrac{-2x}{(x^2-1)^2}$

22. $\dfrac{-8x}{(x^2+1)^2}$ 24. $\dfrac{1}{2\sqrt{x-2}}$ 26. $\dfrac{x}{\sqrt{x^2+1}}$ 28. $\dfrac{1}{2(4-x)^{3/2}}$

30. −1/9   32. 1/6   34. $3x + 2y = -12$ 36. $x - y = -2$

18. $12(x^2-2)^2(3x-4)(2x^2-2x-1)$

20. $(x^3/2)(x+1)^{1/2}(11x+8)$

22. $\dfrac{14x+8}{(3x-1)^{1/3}(4x+5)^{1/2}}$

24. $15(5x-2)^2+4(1-x)^3$

26. $\dfrac{30x^2+50x^4}{(1-5x^2)^3}$

28. $\dfrac{-12x^2+16x+1}{(1-x)^{3/2}}$

30. $\dfrac{3(4x^2-1)^2(4x^2+32x+5)}{(3x+4)^6}$

32. $\dfrac{-68x^3-37x^2+4x+1}{(8x^2+4x)^{3/4}(3x^2+1)^{8/3}}$

34. $\dfrac{20x(1+x^2)^4}{(1-x^2)^6}$

Section 21.8

2. 4/5

4. $-\dfrac{3x}{4y}$

6. $\dfrac{x+4}{y}$

8. $\dfrac{2xy}{y^2-x^2}$

10. $\dfrac{2y-3x^2}{2y-2x}$

12. $\dfrac{-6xy^2-16xy^3-y}{6x^2y+20y^4+24x^2y^2+x}$

14. $\dfrac{(2x^2-3x+2)^4(20x-15)}{4(y+4)^3}$

16. $\dfrac{2x^3+2xy^2-3x}{4y-2x^2y-2y^3}$

18. $3x(x+y)^{1/3}-1$

20. 5/3

22. 3/2

24. $y=4x+4$

26. $2x+5y=12$

28. 16

30. $\dfrac{3}{4}$

Section 21.10

2. $y'=18x^5-24x^2$; $y''=90x^4-48x$; $y'''=360x^3-48$; $y^{iv}=1080x^2$

4. $y'=6x+4$; $y''=6$; $y'''=0$; $y^{iv}=0$

6. $-72/x^5$

8. $(896x^2-64)(4x^2-2)^2$

10. $9(2x+1)^{-5/2}$

12. $\dfrac{80x^2-80}{(x^2+5)^4}$

14. $\dfrac{7}{(x-2)^3}$

16. $y'=\dfrac{3-x}{y}$; $y''=-9/y^3$

18. $y'=\dfrac{-y}{2y+x}$; $y''=\dfrac{2}{(2y+x)^3}$

20. $y'=\dfrac{y^2+y}{x}$; $y''=\dfrac{2y^2(y+1)}{x^2}$

22. $y'=\dfrac{-(y-1)^2}{2}$; $y''=\dfrac{(y-1)^3}{2}$

24. $y'=\dfrac{\sqrt{1+y^2}}{y}$; $y''=-1/y^3$

26. $a=\dfrac{32}{(2t+1)^3}$

28. $a=\dfrac{416}{(5-4t)^3}$

CHAPTER 22

Section 22.1

2.

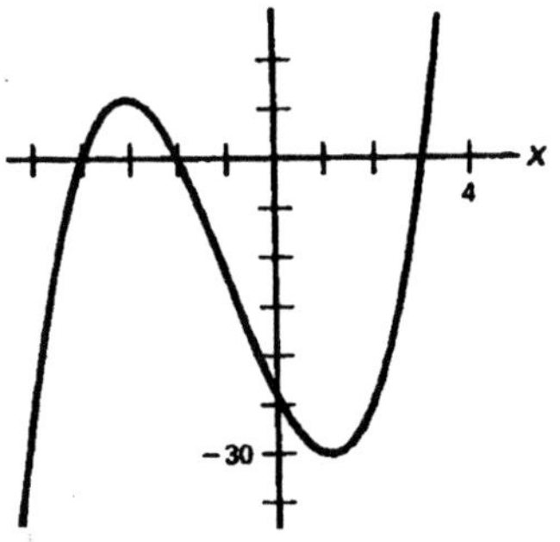

$y = (x + 2)(x - 3)(x + 4)$

4.

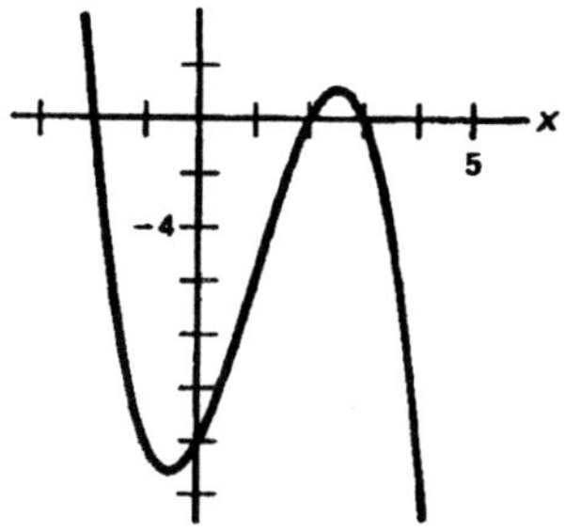

$y = (3 - x)(x^2 - 4)$

6.

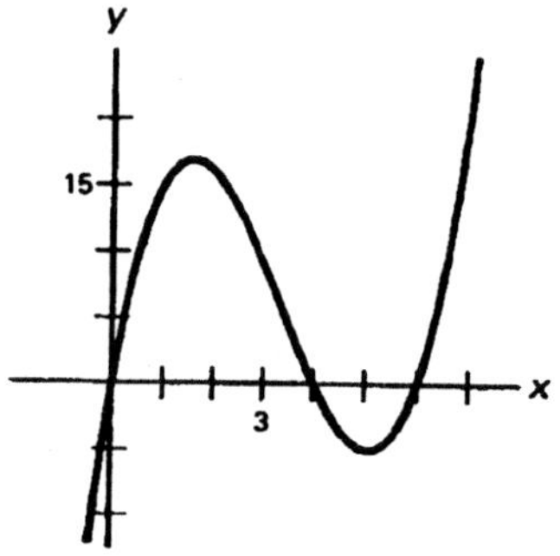

$y = x^3 - 10x^2 + 24x$

8.

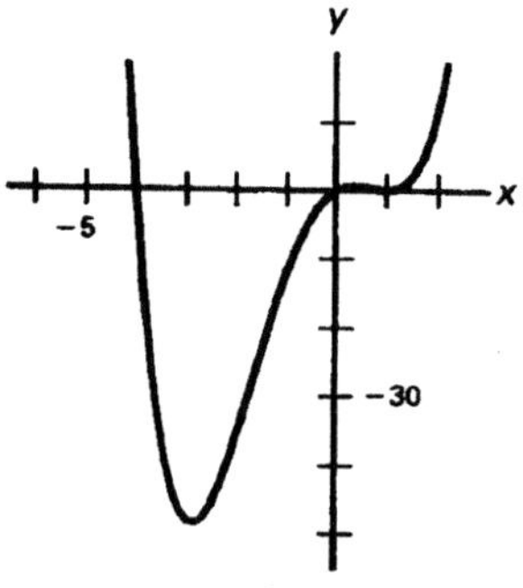

$y = x(x + 4)(x - 1)^2$

10.

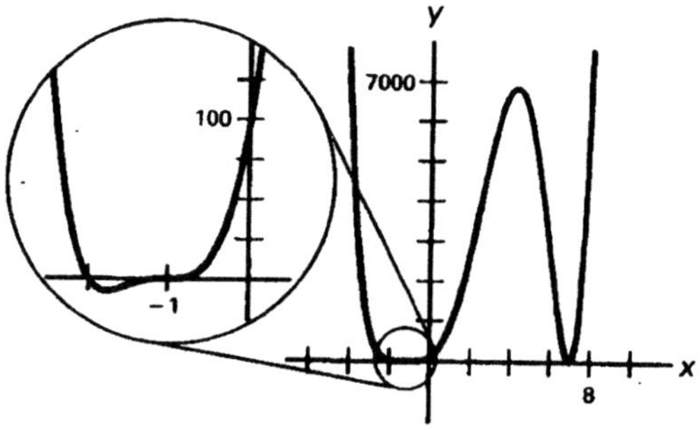

$y = (x + 1)^3(x - 7)^2(x + 2)$

12.

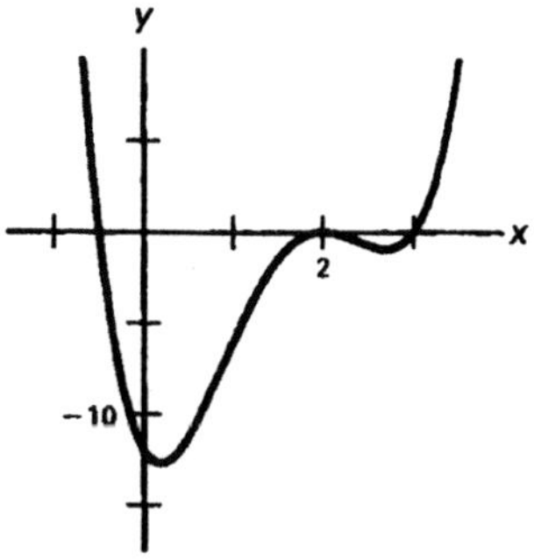

$y = (x - 3)(2x + 1)(x - 2)^2$

14.

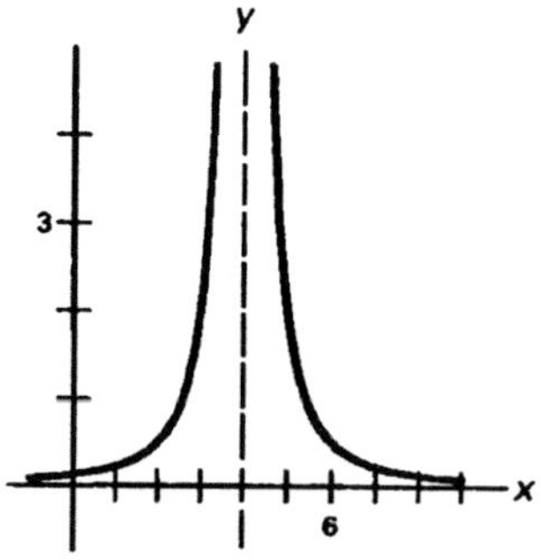

$$y = \frac{2}{(x - 4)^2}$$

16.

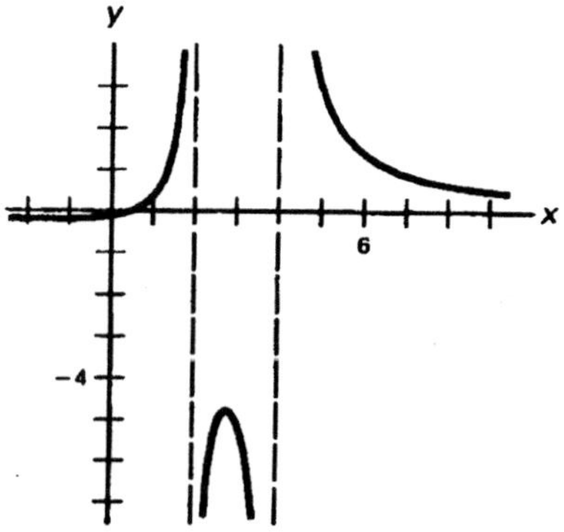

$$y = \frac{2x - 1}{(x - 2)(x - 4)}$$

18.

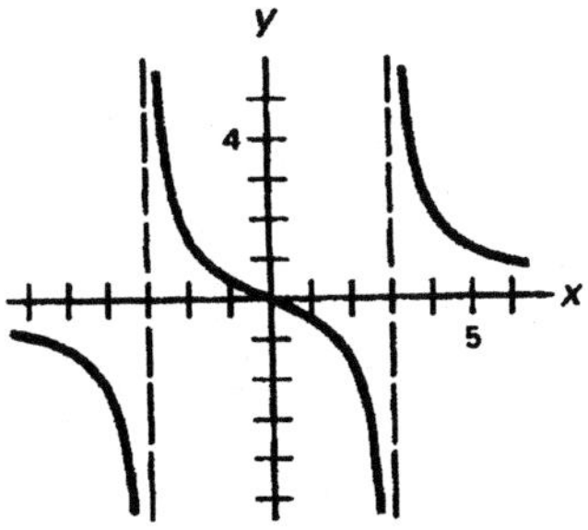

$$y = \frac{4x}{x^2 - 9}$$

20.

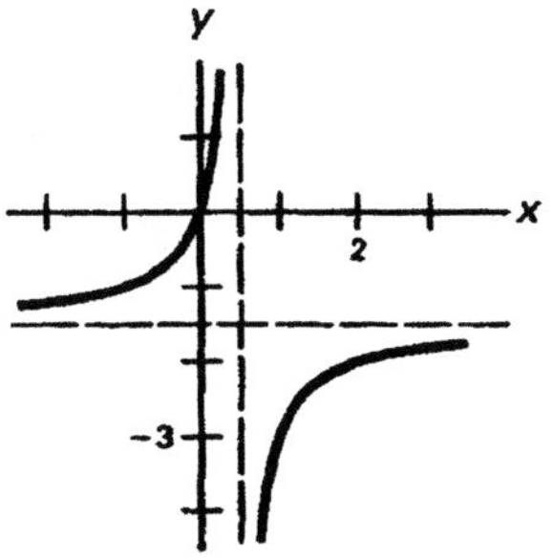

$$y = \frac{3x}{1 - 2x}$$

22.

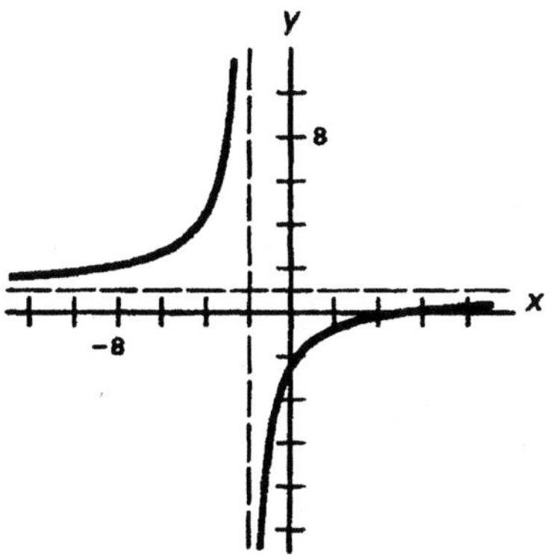

$$y = \frac{x - 5}{x + 2}$$

24.

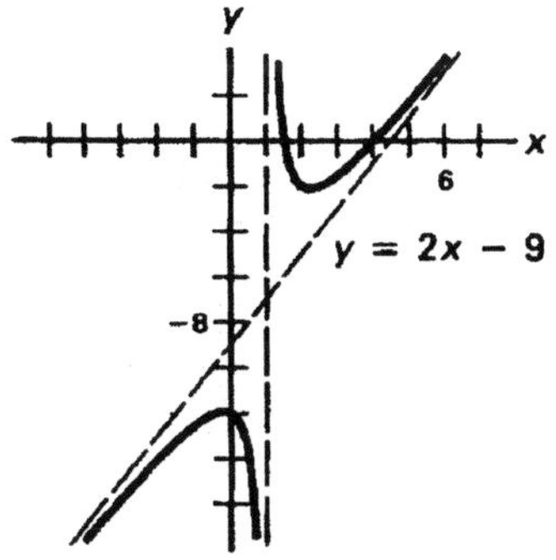

$$y = \frac{2x^2 - 11x + 12}{x - 1}$$

26.

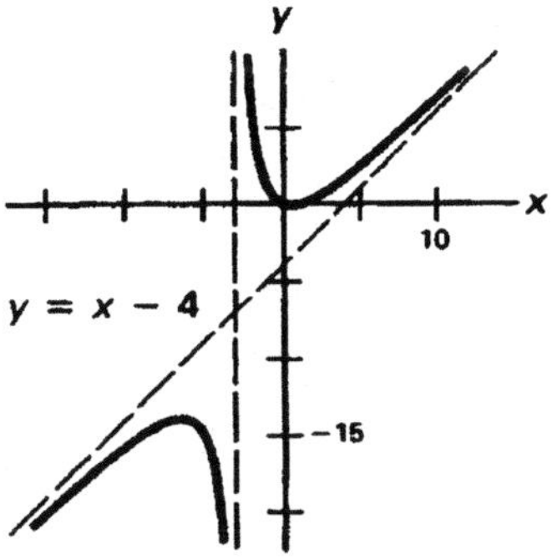

$$y = \frac{x^2 - x}{x + 3}$$

28.

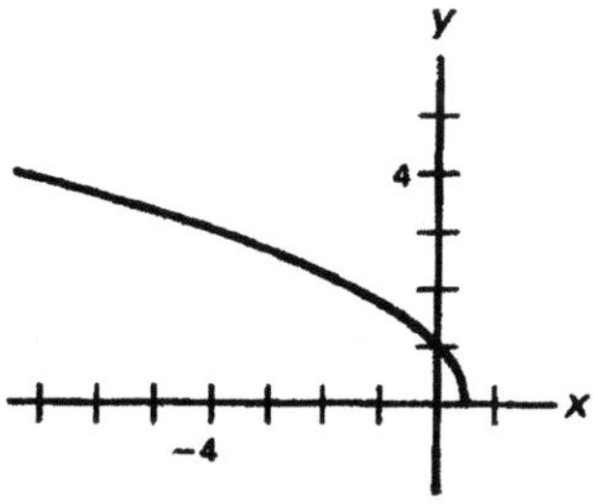

$$y = \sqrt{1 - 2x}$$

30.

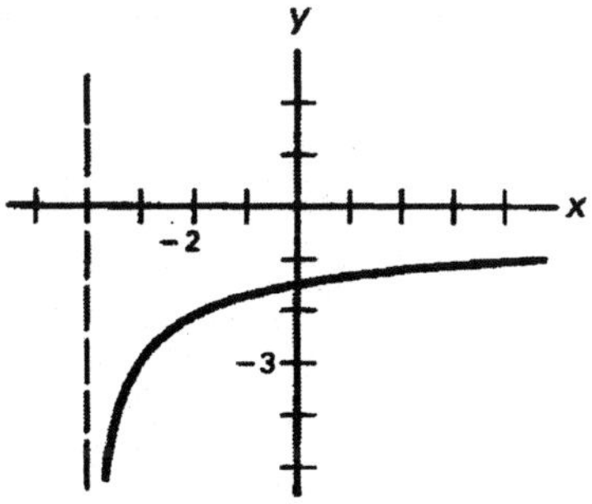

$$y = -\sqrt{\frac{9}{x + 4}}$$

32.

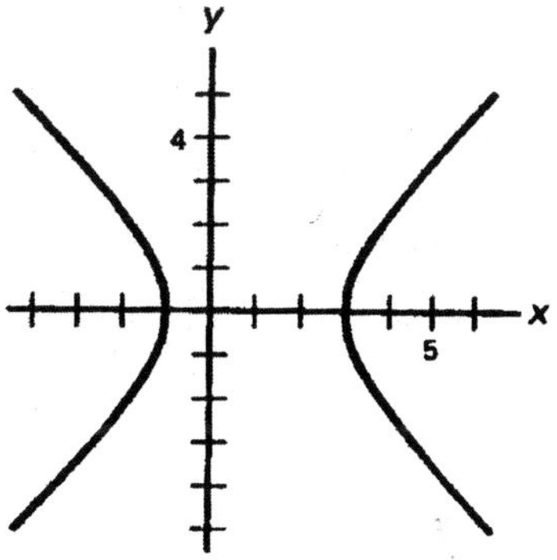

$$y^2 = (x + 1)(x - 3)$$

34.

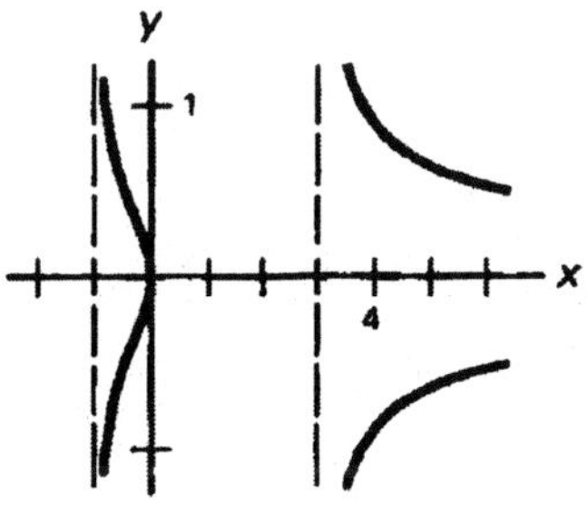

$$y^2 = \frac{x}{(x + 1)(x - 3)}$$

36.

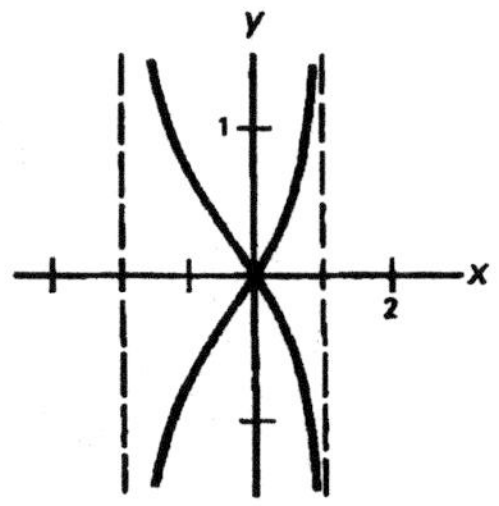

$$y^2 = \frac{x^2}{(1 - x)(x + 2)}$$

Section 22.2

2.

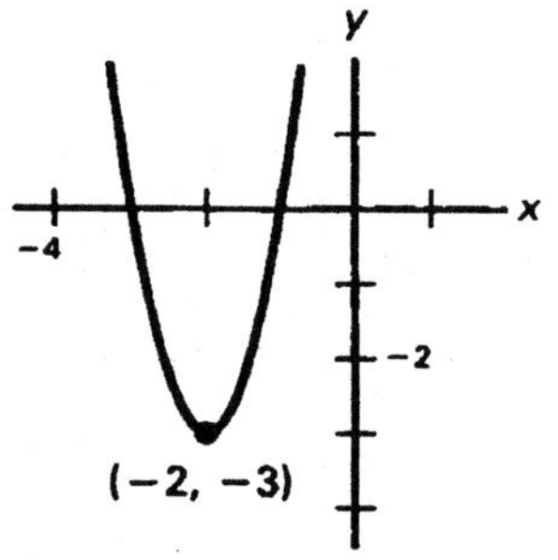

$$y = 3x^2 + 12x + 9$$

4.

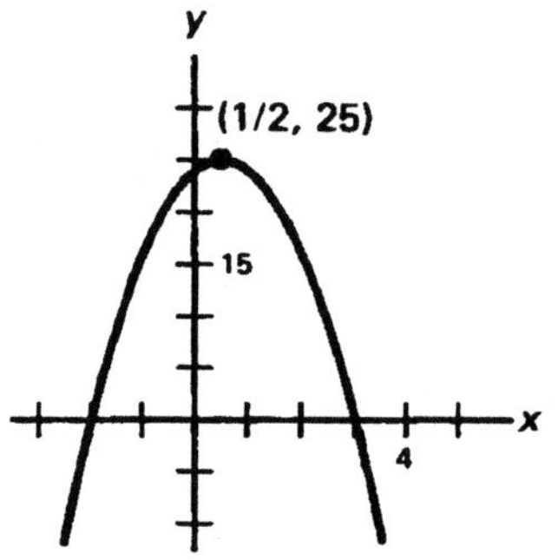

$$y = 24 + 4x - 4x^2$$

6.

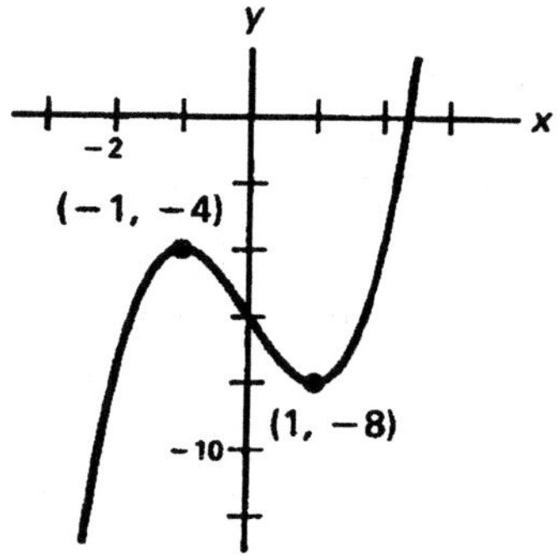

$$y = x^3 - 3x - 6$$

8.

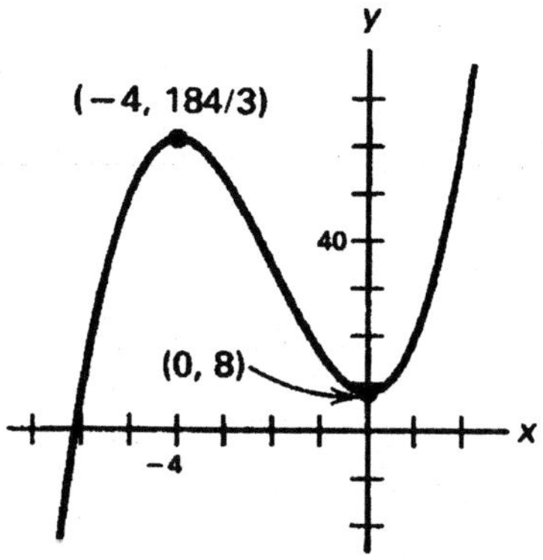

$$y = \frac{5}{3}x^3 + 10x^2 + 8$$

10.

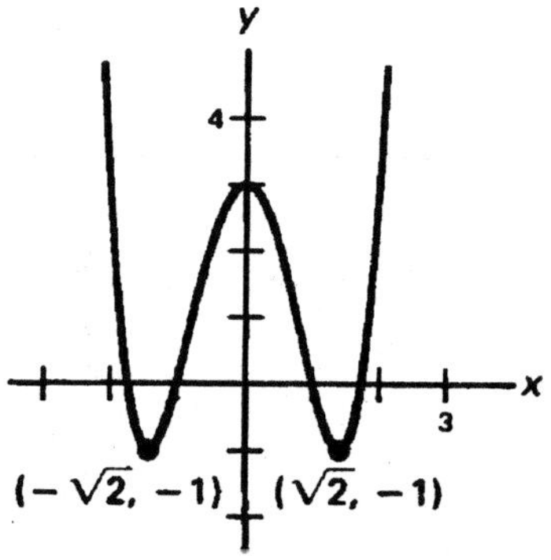

$$y = x^4 - 4x^2 + 3$$

12.

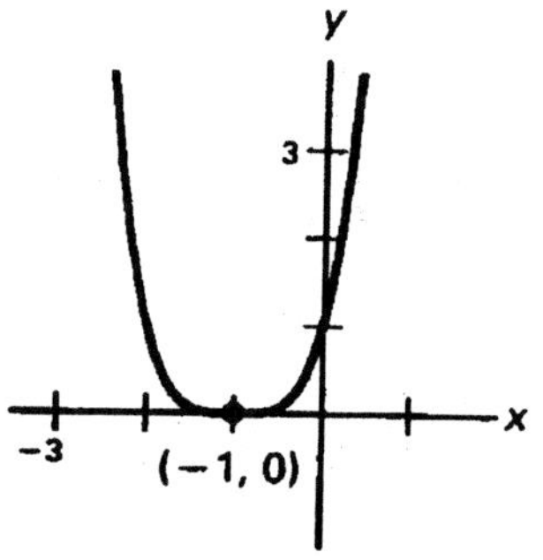

$$y = (x + 1)^4$$

14.

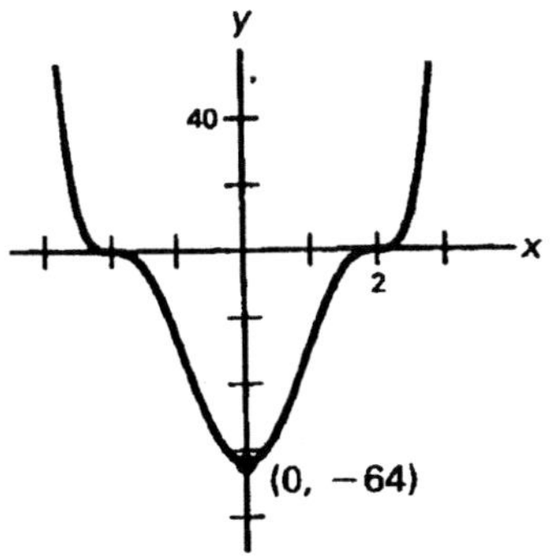

$y = (x^2 - 4)^3$

16.

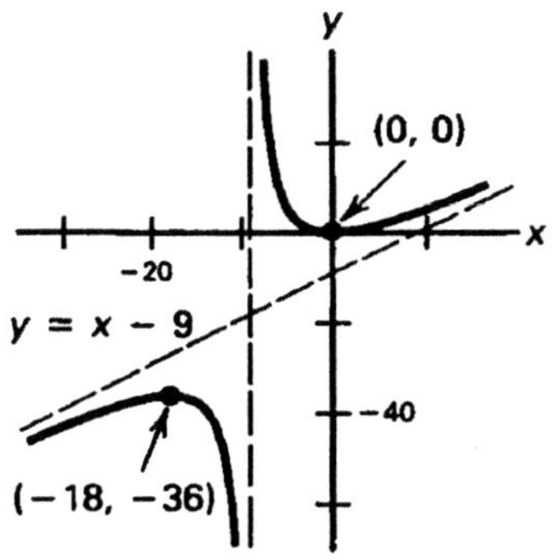

$y = \dfrac{x^2}{x + 9}$

18.

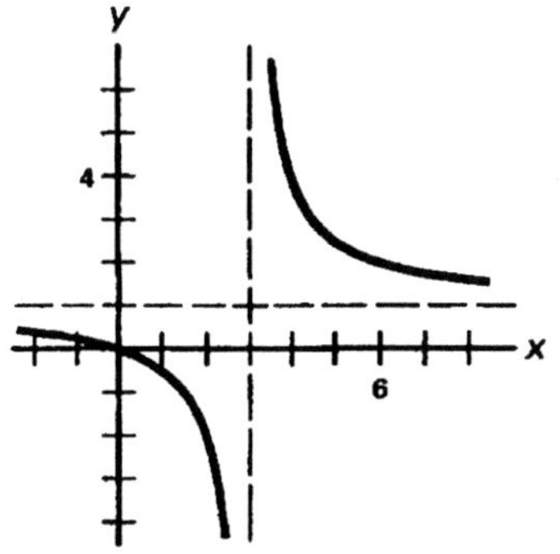

$y = \dfrac{x}{x - 3}$

20.

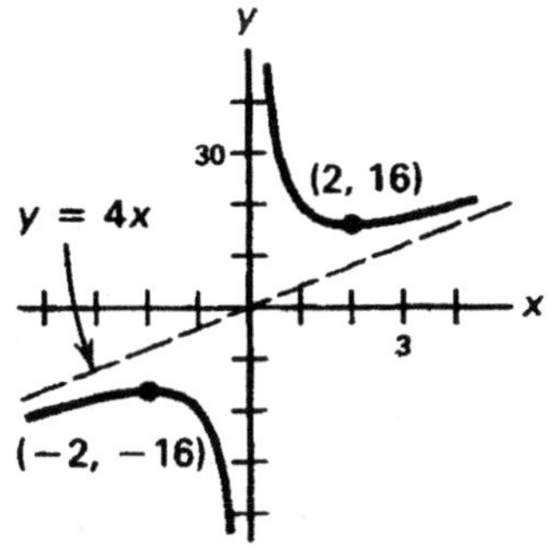

$y = 4x + \dfrac{16}{x}$

22.

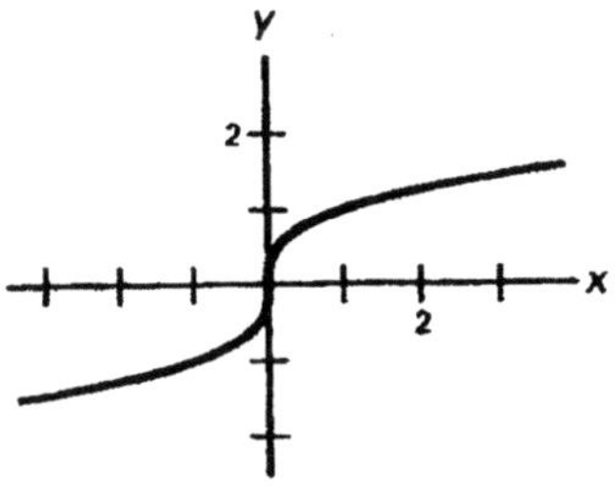

$y = \sqrt[3]{x}$

24.

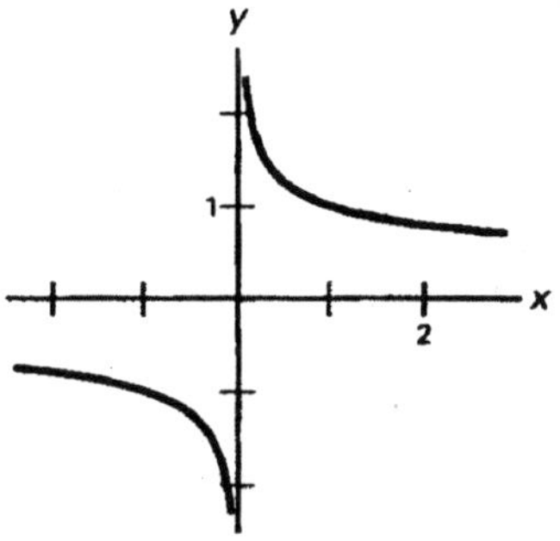

$y = x^{-1/3}$

Section 22.3

2. a) increasing: $x > 1/2$; decreasing: $x < 1/2$
   b) min: $(1/2, -9/4)$
   c) concave upward for all x
   d) no point of inflection

4. a) increasing: $x < 0$; decreasing: $x > 0$
   b) max: $(0, 5)$
   c) concave downward for all x
   d) no point of inflection

2. e) $y = x^2 - x - 2$

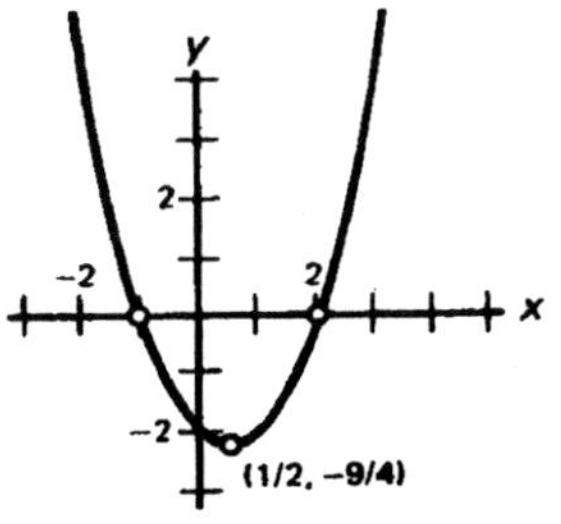

4. e) $y = 5 - x^2$

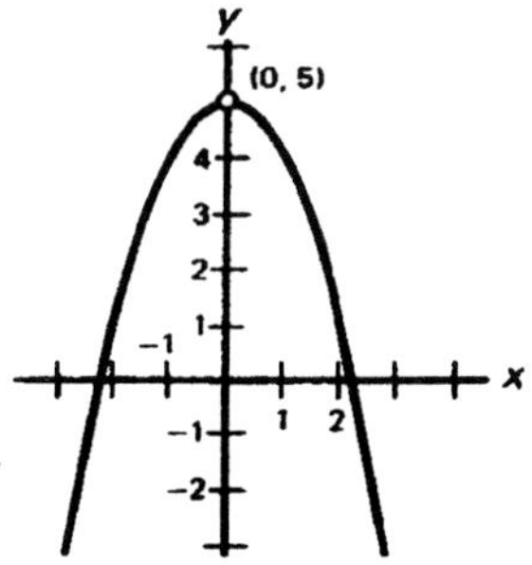

6. a) increasing: $x > 1$; decreasing: $x < 1$ b) min: $(1, -1)$
   c) concave upward for $x < 0, x > 2/3$;
   concave downward for $0 < x < 2/3$
   d) points of inflection: $(0, 0)$ and $(2/3, -16/27)$

8. a) increasing: $-3 < x < 0, x > 3$;
   decreasing: $x < -3, 0 < x < 3$ b) max: $(0, 40)$ min: $(3, -41)$ and $(-3, -41)$
   c) concave upward for $x > \sqrt{3}$ and $x < -\sqrt{3}$;
   concave downward for $-\sqrt{3} < x < \sqrt{3}$
   d) points of inflection: $(\sqrt{3}, -5)$ and $(-\sqrt{3}, -5)$

10. a) increasing for all x b) no max or min
   c) concave upward for $x > 0$; concave downward for $x < 0$;
   d) point of inflection: $(0, 0)$

6. e) $y = 3x^4 - 4x^3$

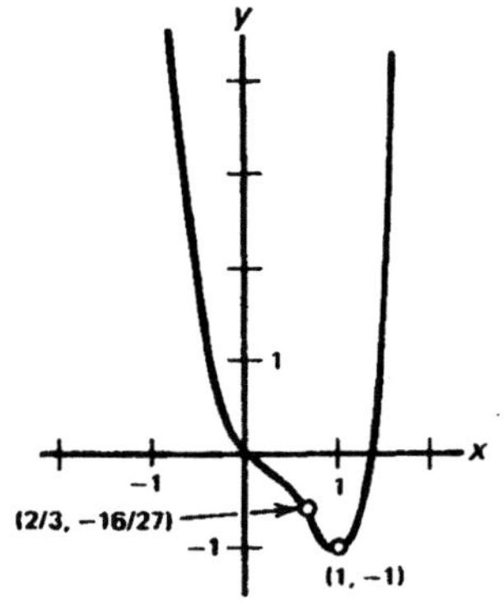

8. e) $y = x^4 - 18x^2 + 40$

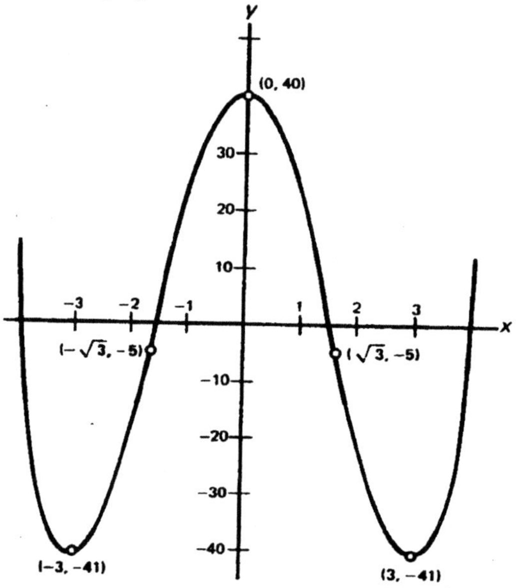

10. e) $y = x^7$

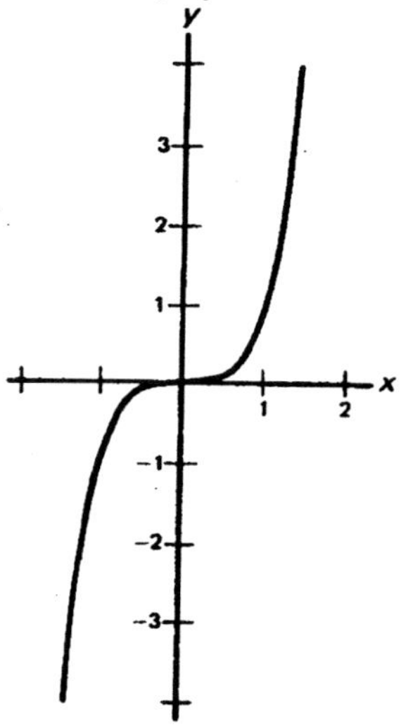

12. a) decreasing: $x < 0$ and $x > 0$ b) no max or min
   c) concave upward for $x > 0$; concave downward for $x < 0$
   d) no point of inflection

14. a) decreasing: $x < 1$ and $x > 1$ b) no max or min
   c) concave upward for $x > 1$; concave downward for $x < 1$
   d) no point of inflection

16. a) increasing: $x < 2$; decreasing: $x > 2$ b) no max or min
   c) concave upward for $x < 2$ and $x > 2$
   d) no point of inflection

12. e) $y = 1/x^3$

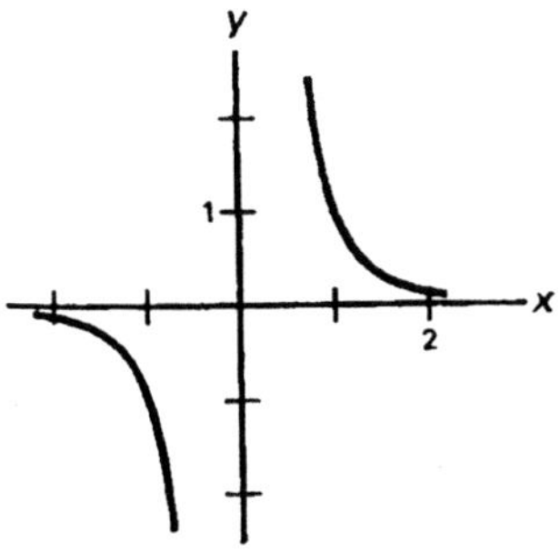

$y = \dfrac{1}{x^3}$

14. e) $y = \dfrac{x}{x-1}$

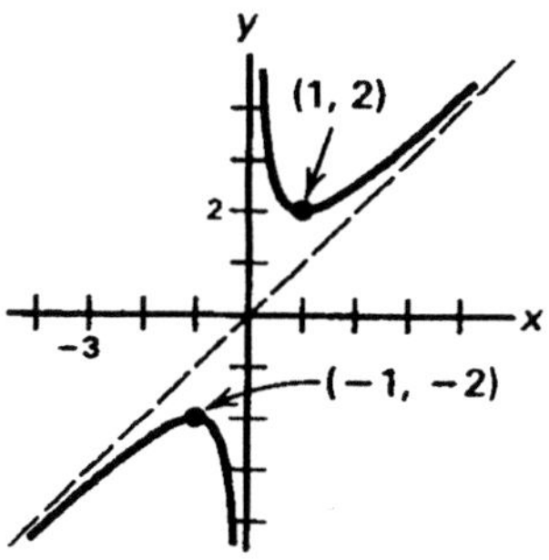

$y = \dfrac{x^2 + 1}{x}$

16. e) $y = \dfrac{1}{(x-2)^2}$

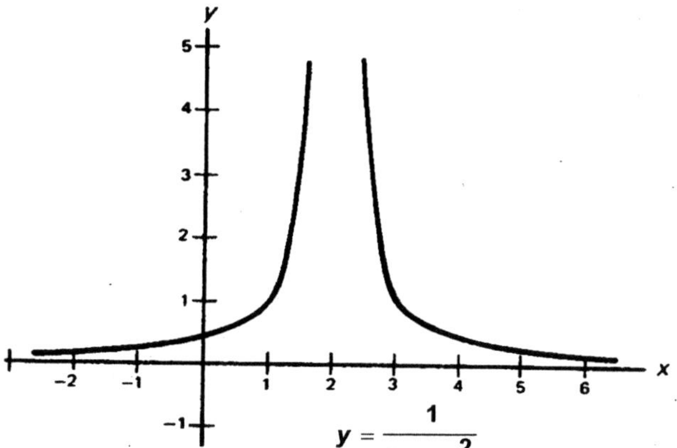

$y = \dfrac{1}{(x-2)^2}$

18. a) increasing: $x < -1, x > 1$; decreasing: $-1 < x < 0, 0 < x < 1$
    b) max: $(-1, -2)$ min: $(1, 2)$
    c) concave upward for $x > 0$, concave downward for $x < 0$
    d) no point of inflection

20. a) decreasing: $x < 2, x > 2$ b) no max or min
    c) concave upward for $x > 2$; concave downward for $x < 2$
    d) no point of inflection

22. a) increasing: $x < -2, -2 < x < 0$: decreasing: $0 < x < 2, x > 2$
    b) max: $(0, -1)$ c) concave upward for $x < -2, x > 2$; concave downward for $-2 < x < 2$
    d) no point of inflection

18. e) $y = \dfrac{x^2 + 1}{x}$

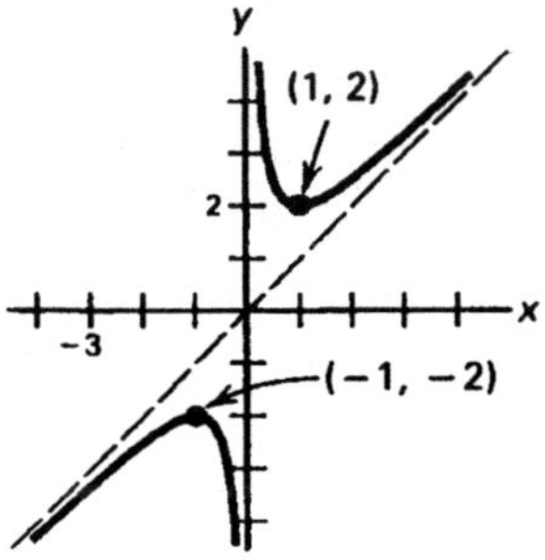

$y = \dfrac{x^2 + 1}{x}$

20. e) $y = \dfrac{x+3}{x-2}$

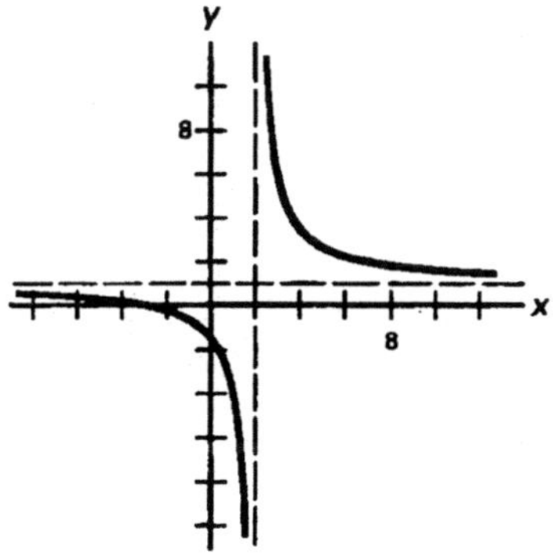

$y = \dfrac{x + 3}{x - 2}$

22. e) $y = \dfrac{4}{x^2 - 4}$

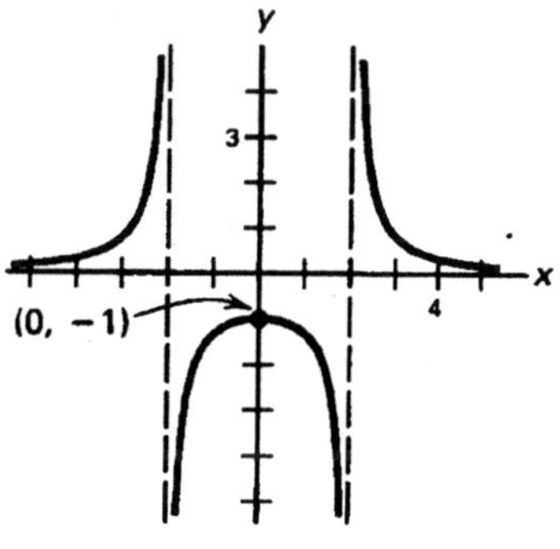

$y = \dfrac{4}{x^2 - 4}$

24. a) increasing: $-8 < x < 0$; decreasing: $x < -8, x > 0$
b) min: $(-8, -1/16)$
c) concave upward for $-12 < x < 0, x > 0$; concave downward for $x < -12$
d) point of inflection: $(-12, -1/18)$
e)

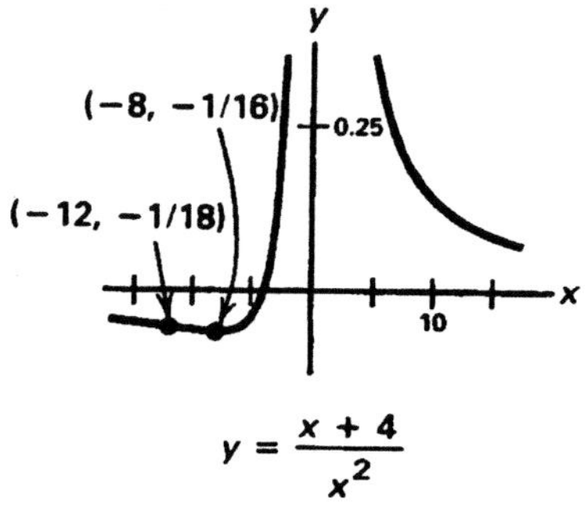

Section 22.4

2. 9.48683 4. 2.71442 6. 1.97435.

8. Estimates -1, 2, 4; better –0.930802, 1.83704, 4.09376

10. Estimates -0.3, 0.7, 1.9; better -0.252567, 0.699705, 1.88620

12. Estimates 0.8, 2.3; better 0.821327, 2.32027

14. Estimates 0.9, 4.9; better 0.844148, 4.91544

16. The numerical answer, 0.343, is given in the problem.

Section 22.5

2. 1 4. 144 in^3 6. 3 in. × 6 in. 8. 3750 ft^2 10. 540 ft

12. $4\sqrt{2} \times 2\sqrt{2}$ 14. 2 C 16. 135 ft × 180 ft

18. $m = -2/3$ at $x = -1/6$ 20. $P = 2/9$ W at $t = 1/3$ s 22. 1.39×10^{-4} A

24. cube: $10\sqrt[3]{4}$ or 15.9 cm on a side 26. $12\pi\sqrt{3}$ or 65.3 in^3 30. $12/\sqrt{7}$ or 4.54 km

Section 22.6

2. 8 4. $3\frac{1}{2}$ 6. 5 mi/h 8. $2\pi\ \text{cm}^2/\text{h}$ 10. 9.42 ft^2/s

12. 0.693 mm/min 14. 2 V/s 16. 16 kPa/min 18. 10/3 ft/s

20. 6.25 W/s 22. –500 cm^3/min 24. 3.89 Ω/s 26. 14,000 W/s

28. −15/16 ft/s (length of shadow is decreasing) 30. –0.990 lb/s

Section 22.7

2. $dy = (20t^4 - 3t^{-1/2})dt$ 4. $ds = \dfrac{11\,dt}{(t+6)^2}$

6. $di = 8t(t^2 + 1)^2(4t^2 + 1)dt$

8. $dm = -3[3t - 1/(3t)]^{-4}[3 + 1/(3t^2)]\, dt$ or $dm = \dfrac{-81t^2(9t^2+1)}{(9t^2-1)^4}dt$

10. $dy = \dfrac{1-4y}{2y+4x}dx$ 12. $dy = -\dfrac{y}{x}dx$ 14. 0.133 16. 2.7 18. 263.9

20. a) 21.60 cm^3
b) 21.69 cm^3
c) 1.25%

22. 1.52 A 24. 2.94 m 26. 8 Ω

CHAPTER 23

Section 23.1

2. $10 \cos 2x$ 4. $-24 \sin 6x$ 6. $12x^3 \sin x^4$ 8. $120x^2 \cos 8x^3$

10. $18 \sin(1 - 3x)$ 12. $(3x^2 + 4x) \cos(x^3 + 2x^2 - 4)$ 14. $-2 \sin(x - 3)$

16. $-6x \sin(x^2 - 4)$ 18. $-4 \sin(1 - 2x) \cos (1 - 2x)$ 20. $-8x \cos^3(x^2 - 1) \sin(x^2 - 1)$

22. $-15(5x + 4)^2 \sin(5x + 4)^3$

24. $24(2x + 1)(x^2 + x - 3)^5 \cos(x^2 + x - 3)^6$

26. $2x \cos 5x \cos x^2 - 5 \sin x^2 \sin 5x$

28. $-4 \sin(3x - 1) \sin(4x + 3) + 3 \cos(4x + 3) \cos (3x - 1)$

30. $5 \sin 3x \cos 5x + 3 \sin 5x \cos 3x$

32. $3(x + 2)^2 \cos(x^2 - 3) \cos (x + 2)^3 - 2x \sin(x + 2)^3 \sin(x^2 - 3)$

34. $-2x\sqrt{4x-3}\sin(x^2+2)+\dfrac{2\cos(x^2+2)}{\sqrt{4x-3}}$

36. $\dfrac{-6x\sin 6x - 2\cos 6x}{3x^3}$ 38. $\dfrac{4\sin(2x-1)-2(4x-5)\cos(2x-1)}{\sin^2(2x-1)}$

40. $6 \cos 3x - 30 \cos 6x$ 42. $2x - 18 \sin^2 2x \cos 2x$ 44. $-\csc^2 x$

46. $-\sin x$ 48. $\cos x$ 50. $-18 \sin 3x + 180 \sin 6x$ 52. 4

54. $12x + y = 5\pi/2 - 3\sqrt{3}$

Section 23.2

2. $-2 \csc^2 2x$ 4. $-4 \csc 4x \cot 4x$ 6. $3x^2 \sec^2(x^3 + 4)$

8. $4 \sec(x - \pi/3) \tan(x - \pi/3)$ 10. $28 \sec^4 7x \tan 7x$

12. $-12x^3 \csc^3 x^4 \cot x^4$ 14. $(5 \sec^2 5x)/(2\sqrt{\tan 5x})$

16. $(-2x \csc^2 2x - \cot 2x)/x^2$ 18. $2 \sec^2 2x - 3 \csc^2 3x$ 20. $\sec^2 x$

22. $\sin x \tan x(1 + 2 \sec^2 x)$ 24. $2x \tan x(x \sec^2 x + \tan x)$

26. $2x \cot x(\cot x - x \csc^2 x)$ 28. $-4 \sin 4x \cot 2x - 2 \cos 4x \csc^2 2x$

30. $-2 \csc^2 x(\cos 2x \cot x + \sin 2x)$ 32. 0

34. $3(\sin x - \tan^2 x)^2(\cos x - 2\tan x \sec^2 x)$

36. $-12\cot^2 x \csc^2 x(1 + \cot^3 x)^3$

38. $-2\sec(2\cos x)\tan(2\cos x)\sin x$

40. $\csc^2 x \sin(\cot x)$

42. $2\cos x \tan(\sin x)\sec^2(\sin x)$

44. $\dfrac{1+\sin^2 x}{\cos^3 x}$

46. $-\dfrac{1+2\cos^2 x}{\sin^4 x}$

48. $\dfrac{1}{1+\cos x}$

50. $4\sec 2x(2\sec^2 2x - 1)$

52. $\dfrac{2x^2\sec^2 x\tan x - 2x\sec^2 x + 2\tan x}{x^3}$

54. 4

Section 23.3

2. $\dfrac{-6}{\sqrt{1-36x^2}}$

4. $\dfrac{1}{|x|\sqrt{16x^2-1}}$

6. $\dfrac{-1}{2\sqrt{x}(1+x)}$

8. $\dfrac{-4}{|x|\sqrt{x^2-1}}$

10. $\dfrac{6x}{x^4-2x^2+2}$

12. $\dfrac{-12}{|x|\sqrt{x^2-4}}$

14. $\dfrac{10\text{ Arctan }5x}{1+25x^2}$

16. $\dfrac{12\text{Arcsec}^2 2x}{|x|\sqrt{4x^2-1}}$

18. $\dfrac{-12\text{Arcsin}(1-x)}{\sqrt{2x-x^2}}$

20. $\dfrac{x^2}{1+x^2}$

22. $\dfrac{-x-1}{\sqrt{1-x^2}}$

24. $\dfrac{-x^2}{\sqrt{1-x^2}} + 2x\text{ Arccos }x$

26. $\dfrac{2x^2}{\sqrt{1-x^4}} + \text{Arcsin }x^2$

28. $\dfrac{x^2}{(1-x^2)^{3/2}}$

30. $\dfrac{x-(1+x^2)\text{Arctan }x}{x^2(1+x^2)}$

32. 1/2

34. $-2\pi/3$

Section 23.4

2. $\dfrac{2x\log e}{x^2-1}$

4. $\dfrac{5\log_7 e}{5x+2}$

6. $\dfrac{2x}{x^2-4}$

8. $2x\tan x^2$

10. $\dfrac{3\cos 2x - 2x\sin 2x}{x\cos 2x}$ or $\dfrac{3}{x} - 2\tan 2x$

12. $\dfrac{2x}{3(x^2-1)}$

14. $\dfrac{1+x^2}{x(1-x^2)}$

16. $(\ln x)(\cos x) + \dfrac{\sin x}{x}$

18. $2x+\dfrac{3}{3x-2}$ or $\dfrac{6x^2-4x+3}{3x-2}$

20. $\dfrac{1}{x\sqrt{1-\ln^2 x}}$

22. $\dfrac{3}{(\text{Arctan } x)(1+x^2)}$

24. $(1/2)\sqrt{x(2x+1)(1-5x)}\left\{\dfrac{1}{x}+\dfrac{2}{2x+1}-\dfrac{5}{1-5x}\right\}$

26. $\dfrac{x(2x-1)^{3/2}}{\sqrt[3]{x+1}}\left\{\dfrac{1}{x}+\dfrac{3}{2x-1}-\dfrac{1}{3(x+1)}\right\}$

28. $x^{x+1}\left\{\dfrac{x+1}{x}+\ln x\right\}$

30. $(1+x)^{1/x}\left\{\dfrac{1}{x(1+x)}-\dfrac{\ln(1+x)}{x^2}\right\}$

32. $(\tan x)^x\left\{\dfrac{x}{\sin x\cos x}+\ln(\tan x)\right\}$

34. $x^{\sin x}\left\{\dfrac{\sin x}{x}+(\ln x)(\cos x)\right\}$

36. $y=2x-2$

38. $y=x-1$

Section 23.5

2. $6e^{3x+1}$

4. $(2x+3)e^{x^2+3x}$

6. $\dfrac{2x(4^{x^2})}{\log_4 e}$

8. $-3x^2e^{-x^3}$

10. $\dfrac{3e\sqrt{3x-2}}{2\sqrt{3x-2}}$

12. $-10x(\sin x^2)e^{\cos x^2}$

14. $(3x^3+1)e^{x^3}$

16. $6x(\sec x^2)e^{\tan x^2}$

18. $4e^{4x}\sec^2 e^{4x}$

20. $\dfrac{e^x(x-1)}{x^2}$

22. $\dfrac{(2x^2-4x-1)e^{x^2}}{(x-2)^2}$

24. $(4e^{4x})/(1-e^{8x})^{1/2}$

26. $\dfrac{2xe^{x^2}}{1+e^{2x^2}}$

28. $3(e^x+e^{-x})(e^x-e^{-x})$

30. $\dfrac{1-e^x}{x-e^x}$

Section 23.6

2. 1   4. 3   6. $\frac{1}{4}$ 8. $\frac{1}{3}$ 10. 0   12. 3   14. Does not exist

16. 0   18. 1   20. $\frac{1}{2}$ 22. y = 3

Section 23.7

2. $y = x \ln x, x > 0$

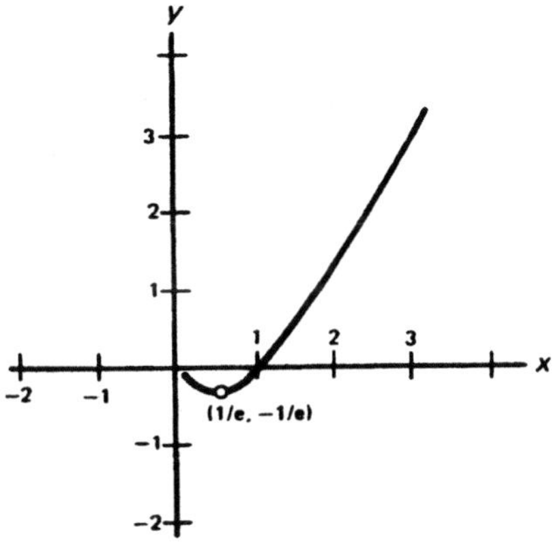

4. $y = \ln(\sin x)$, $0 < x < \pi$

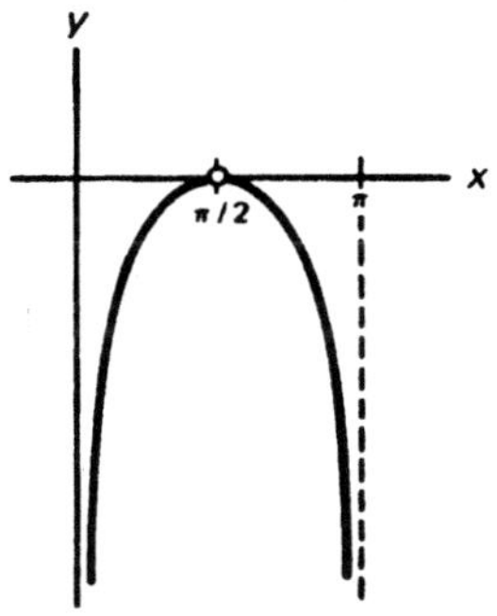

6. $y = \ln x - x$, $x > 0$

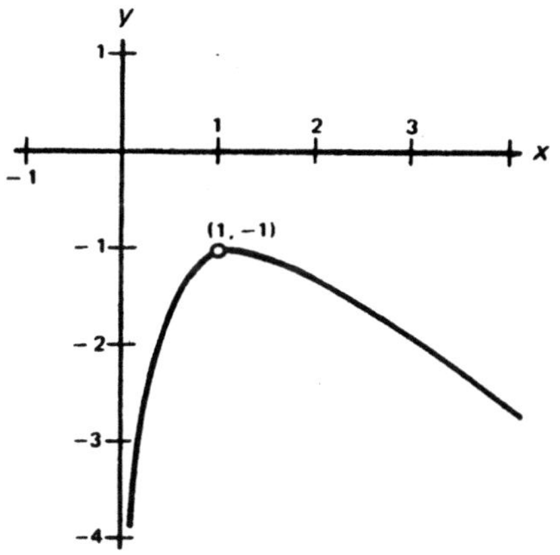

8. $y = \ln\left(\frac{1}{x^2 - 1}\right)$

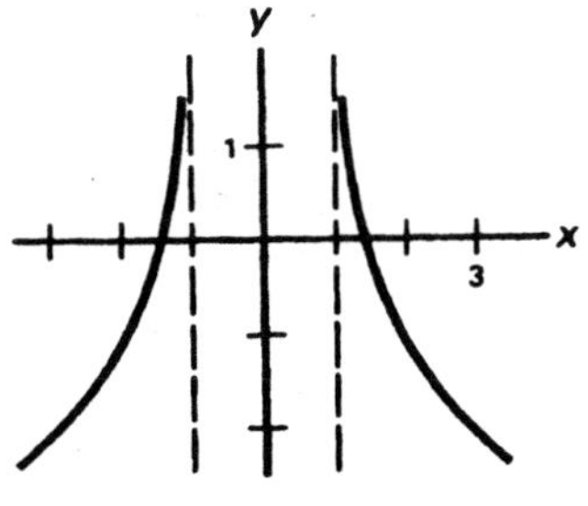

$y = \ln\left(\frac{1}{x^2 - 1}\right)$

10. $y = \frac{x}{\ln x}, x > 0$

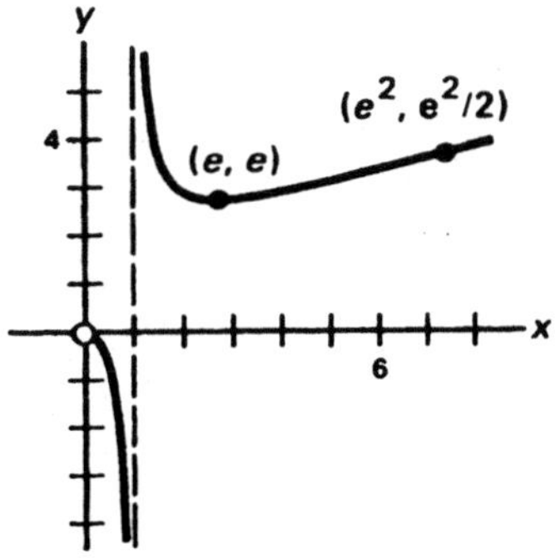

$y = \frac{x}{\ln x}, x > 0$

12 max: $(-1, e^{-2})$; min: $(0, 0)$

14. min: $(1, e)$

16. min: $(e^2, e^2/4)$

18. $y = 2ex - e$

20. $y = -x + 1 + \pi/2$

22. $-4e^{-x} \cos x$

24. $V_L = 8000e^{-20t}$

26. 4

28. $p = 50 \sin t \cos t$ or $25 \sin 2t$

30. $-2 \times 10^{-8}\, e^{-0.001t}$

32. $v = \frac{2t}{t^2+1}$; $a = \frac{2-2t^2}{(t^2+1)^2}$

34. $v = 6.08$; $a = 10.4$

CHAPTER 24

Section 24.1

2. $(1/25)x^{25} + C$ 4. $20x^{10} + C$ 6. $(16/5)x^{5/2} + C$ 8. $-3/x + C$

10. $6x^{1/6} + C$ 12. $(7/11)x^{11} + (8/5)x^5 - (11/4)x^4 + C$ 14. $(1/2)x^4 + (3/2)x^2 + 5/(3x^3) + C$

16. $(1/5)x^5 - (10/3)x^3 + 25x + C$ 18. $(3/32)(8x - 1)^{4/3} + C$

20. $(x^3 + 2)^6 + C$ 22. $(1/6)(6x^2 + 5)^{3/2} + C$

24. $(1/40)(x^5 + 3)^8 + C$ 26. $3\sqrt{8x^2 - 1} + C$

28. $(1/5)(x^4 - 3x^2)^5 + C$ 30. $(-1/2)(3x^2 + x - 2)^{-2} + C$

32. $(2/3)(x^3 - x^2 + 2)^{3/2} + C$ 34. $\sqrt{x^2 - 8x + 3} + C$

36. $(1/12)(4x - 5)^3 + C$ 38. $(1/35)(5x + 3)^7 + C$

40. $x^5/5 - 8x^3/3 + 16x + C$ 42. $(x^2 - 4)^3 + C$

44. $(1/12)(6x^4 - 1)^6 + C$ 46. $6\sqrt{2x^4 - 1} + C$

48. $4(5x^4 + 6x^2)^{3/4} + C$ 50. $-2/(9x^3) + 1/(2x) + C$

Section 24.2

2. $y = 5x^3/3 - 2$ 4. $y = x^4 - x^2 + 2x - 4$ 6. $y = \sqrt{x^2 - 1}$

8. $s = 2t^3/3 - t^2 - 65t + 10$ 10. a) $h = s = -4.90t^2 + 500$ b) 10.1 *s*

12. 25 cm/s^2 14. a) 25 ft   b) 2.5 s   c) 40 ft/s

16. a) $s = -4.90t^2 - 10t + 80$ b) 3.15 s   18. 14.7 J   20. 230 V   22. 44.8 *C*

Section 24.3

2. 75/2   4. 26/3   6. 3   8. 5/72   10. $9 - \sqrt{3}$ or 7.27   12. 112

14. 256/5   16. 16/9   18. 43.6   20. 13/27   22. 2/81   24. 32

26. 125/24   28. 64/3   30. 324/5

Section 24.4

2. 1/5   4. 60.   6. 37/5   8. 0   10. 663   12. 112/5   14. 182/3

16. 777/4   18. 62/5   20. −8/3   22. $(20\sqrt{10} - 4\sqrt{2})/9$ 24. 0   26. $2\sqrt{10} - 4$

CHAPTER 25

Section 25.1

2. 7/3 4. 3 6. 32/3 8. 65/12 10. 2/3 12. 17/12 14. 2

16. 9/2 18. 64/3 20. 9 22. 1/3 24. 3 26. 253/12

Section 25.2

2. $56\pi/3$ 4. $15\pi/2$ 6. $8\pi/5$ 8. $5\pi/2$ 10. $8\pi/3$

12. $38\pi/15$ 14. $128\pi/5$ 16. $4\pi/21$ 18. $38\pi/15$ 20. $64\pi/15$

22. $16\pi/3$ 24. $104\pi/5$ 26. 100π 28. $76\pi\ \text{in}^3$

Section 25.3

2. $64\pi/5$ 4. $128\pi/5$ 6. $64\pi/5$ 8. 8π 10. $27\pi/2$

12. $8\pi/3$ 14. $1048\pi/15$ 16. $4\pi/15$ 18. 8π 20. $32\pi/15$

Section 25.4

2. −3/2 4. 2.51 6. −6 8. −2 10. 55

12. 288 14. 4.52 mi from Leadville 16. (−10/3, −3) 18. (2.68, −1.18)

20. (13, −36.5) 22. 6 24. 1.5 mi east and and 6.23 mi south of A

Section 25.5

4. 6 cm 6. 2.7 cm 8. 3 cm from given end 10. (4, 5.29)

12. (7.91, 4.89) 14. (0, 3.76) 16. (0, 12/5) 18. (0, 0)

20. (−1/2, 12/5) 22. (2/5, 4/7) 24. $(4/(3\pi), 4/(3\pi))$ 26. (0, 5/8)

28. (0, 3/4) 30. (3/8, 0)

Section 25.6

2. 725; 6.02 4. 4068; 6.22 6. 1280; 5.80 8. 4320, 6.41

10. 1280/7; 2.62 12. 1/5; 0.548 14. 162; 1.34 16. 768; 2.19

18. 3888π; 3.29 20. $256\pi/3$; 1.63 22. 64π; 1.15 24. 7015; 5.48

Section 25.7

2. 666
4. 90 in.-lb
6. 750 N·cm or 7.5 J
8. 1.695×10^{-16} J
10. 112,500 ft-lb
12. 56,500 ft-lb
14. 828,100 ft-lb
16. a) 23,960 ft-lb b) 6533 ft-lb
18. 21,000 lb
20. 176,400 N
22. 1279.2 lb
24. 1490 N
26. 52,416 lb
28. 19,169 lb
30. 14/9
32. 4
34. 5.95 W

CHAPTER 26

Section 26.1

2. $(3/8)(2x - 1)^{4/3} + C$

4. $\dfrac{-1}{4(2x+1)^2} + C$

6. $(1/3)(x^2 + 4)^{3/2} + C$

8. $\sqrt{\sin 2x} + C$

10. $(-1/2)\cos^2 x + C$ or $(1/2)\sin^2 x + C$

12. $(-1/6)(1 - \sin 2x)^3 + C$

14. $(8 - \cot 2x)^{1/2} + C$

16. $(1/8)(10 - e^{-2x})^4 + C$

18. $2(2 + e^{\tan x})^{1/2} + C$

20. $(1/2)\ln^2|4x| + C$

22. $(1/4)\ln^2(x^2 + 1) + C$

24. $(1/4)\ \text{Arctan}^2 2x + C$

26. $\sqrt{9 + \tan x^2} + C$

28. $(1/6)\ln^3|4x| + C$

30. $37^{3/2} - 1$ or 224.1

32. 1/24

34. $\pi^2/64$

36. 1/3

Section 26.2

2. $\ln|x + 5| + C$

4. $(1/4)\ln|4x + 2| + C$

6. $(3/2)\ln|x^2 + 1| + C$

8. $(1/2)\ln|x^2 + 2x - 3| + C$

10. $-\ln|\cos x| + C$ or $\ln|\sec x| + C$

12. $(1/2)\ln|1 + \sin 2x| + C$

14. $(1/4)\ln|1 - \cot 4x| + C$

16. $(1/2)\ln|x^2 + \cos 2x| + C$

18. $\ln|4 + \ln|x|| + C$

20. $(1/3)e^{3x-1} + C$

22. $\dfrac{1}{2e^{2x+3}} + C$

24. $(1/4)e^{x^4-1} + C$

26. $(-1/2)e^{-x^2} + C$

28. $(1/2)e^{2\tan x} + C$

30. $e - 1$ or 1.72

32. $\dfrac{1}{2}\dfrac{10^{2x}}{\ln 10} + C$

34. $\ln|1 - e^{-x}| + C$

36. $\ln\sqrt{2}$ or $(1/2)\ln 2$ or 0.347

38. $(1/5)(\ln 34 - \ln 4)$ or 0.428

40. $(-1/4)(1 - e^8)$ or 745.0

42. $(1/2)(e^4 - e)$ or 25.9   44. ln (3/2) or 0.405   46. ln 3 or 1.10

48. $1 - e^{-5}$ or 0.993

Section 26.3

2. $(1/6) \sin 6x + C$

4. $(-1/2) \cos (2x + 7) + C$

6. $(1/4) \sin x^4 + C$

8. $(-1/2) \cos (x^2 - 2x + 3) + C$

10. $(1/2) \tan 2x + C$

12. $(-1/7) \csc 7x + C$

14. $(-1/3) \cot (3x - 2) + C$

16. $2 \sec(x/2) + C$

18. $(-1/3) \cot x^3 + C$

20. $(1/6) \sec 3x^2 + C$

22. $(1/2) \ln |\sin x^2| + C$

24. $(1/30 \ln |\csc x^3 - \cot x^3| + C$

26. $\ln |\sec (\ln |x|) + \tan (\ln |x|) + C$

28. $(-2/3) \ln|\cos 3x| + (1/3) \tan 3x + C$

30. $\ln|\sec x + \tan x| - \ln |\sin x| + C$ 32. 2/3   34. 0   36. 1

38. $\dfrac{\sqrt{2}-1}{\pi}$ 40. 2   42. $\sqrt{2}-1$ 44. $1 + \sqrt{2}$

Section 26.4

2. $\sin x - (1/3) \sin^3 x + C$

4. $-\cos x + \cos^3 x - (3/5) \cos^5 x + (1/7) \cos^7 x + C$

6. $(-1/3) \cos^3 x + C$ 8. $(1/4) \sin^4 x + C$

10. $(1/9) \sin^3 x - (1/15) \sin^5 3x + C$ 12. $x/2 + (1/16) \sin 8x + C$

14. $3x/8 - (1/28) \sin 14x + (1/224) \sin 28x + C$

16. $x/16 - (1/64) \sin 4x - (1/48) \sin^3 2x + C$

18. $3x/128 - (1/256) \sin 8x + (1/2048) \sin 16x + C$

20. $(-1/16) \cot^4 4x + (1/4) \cot^2 4x + (1/4) \ln|\sin 4x| + C$

22. $(1/21) \tan^3 7x + (1/7) \tan 7x + C$

24. $(-1/3) \cot 3x - (1/9) \cot^3 3x + C$

26. $(-1/5)\cot^5 x + (1/3)\cot^3 x - \cot x - x + C$

28. $\dfrac{5\pi^2}{32}$

Section 26.5

2. $\text{Arcsin}\, x + C$

4. $(1/5)\,\text{Arcsin}\,(5x/12) + C$

6. $(1/2)\,\text{Arctan}\,(x/2) + C$

8. $(1/20)\,\text{Arctan}\,(5x/4) + C$

10. $((1/\sqrt{6})\ \text{Arcsin}\ [\sqrt{(6/5)}x] + C$

12. $1/(2\sqrt{3})\ \text{Arcsin}\ (2x/\sqrt{3}) + C$

14. $(1/4)\,\text{Arctan}[(x-3)/4)] + C$

16. $(1/3)\,\text{Arctan}\,[(x+2)/3] + C$

18. $\text{Arctan}\,(\sec x) + C$

20. $\text{Arcsin}[(1/2 \sin x] + C$

22. $\pi/2$

24. 0.0687

26. $s = \text{Arcsin}\,(t/3) + 10$

Section 26.6

2. $\ln\left|\dfrac{x-3}{x-2}\right| + C$

4. $\ln\left|\dfrac{x}{x+1}\right| + C$

6. $\ln\left|\dfrac{x-2}{x-1}\right| + C$

8. $(-10/3)\ln|2+x| - (1/3)\ln|1-x| + C$ or $\dfrac{1}{3}\ \ln\left|\dfrac{x-1}{(x+2)^{10}}\right| + C$

10. $2\ \ln\left|\dfrac{x-1}{x+2}\right| - \dfrac{1}{x-1} + C$

12. $\ln\left|\dfrac{(x+2)^2}{x}\right| - \dfrac{1}{x} + C$

14. $x - 2\ln|x+1| - \dfrac{1}{x+1} + C$

16. $\ln|x^2+4| + (1/2)\,\text{Arctan}\,(x/2) + 3\ln|x-1| + C$

18. $(1/2)\ln|(x^2+1)(x^2+2)| + C$

20. $\text{Arctan}\, x + \dfrac{1}{x^2+1} + C$

22. $(1/6)[5\ln(9/7) + \ln 3]$ or 0.393

24. $-4/3 + 3\ln(3/2)$ or -0.117

26. $-1/6 + \ln(3/2)$ or 0.239

Section 26.7

2. $x\sin x + \cos x + C$

4. $-xe^{-x} - e^{-x} + C$

6. $(1/3)x^3\ln|x| - (1/9)x^3 + C$

8. $2\sqrt{x}\,(\ln|x| - 2) + C$

10. x Arctan $x - (1/2) \ln |x^2 + 1| + C$

12. $-x^2 \cos x + 2x \sin x + 2 \cos x + C$

14. $(1/4) x^4 \ln|x| - x^4/16 + C$ 16. $(1/4)e^{2x}(2x^2 - 2x + 1) + C$

18. $(1/2)(x^2 + 1)$ Arctan $x - x/2 + C$

20. $(1/27) e^{3x} (9x^3 - 9x^2 + 6x - 2) + C$

22. $\pi/2 - 1$ 24. 116/15

26. $\pi^2/4 - 2$ 28. $2(e^2 - 1)$ 30. $2\pi^2$

Section 26.8

2. $(1/4)$ Arcsin $(4x/3) + C$

4. $(-x/32)\sqrt{1-16x^2} + (1/128)$ Arcsin $4x + C$ 6. $\ln|2 + 2\sqrt{5}| - \ln 4$

8. $(1/2)[\ln|6 + 3\sqrt{3}| - \ln|4 + \sqrt{7}|]$ or $\frac{1}{2}\ln\left|\frac{6+3\sqrt{3}}{4+\sqrt{7}}\right|$

10. $\frac{1}{3}\ln\left|\frac{3-\sqrt{9-x^2}}{x}\right| + C$ 12. $\frac{x}{4\sqrt{x^2+4}} + C$

14. $-\frac{1}{3}\ln\left|\frac{3+\sqrt{x^2+9}}{x}\right| + C$ 16. $\sqrt{4+9x^2} - 2\ln\left|\frac{\sqrt{4+9x^2}+2}{3x}\right| + C$

18. $\frac{-x}{5\sqrt{x^2-5}} + C$ 20. $\ln\left|\frac{1-\sqrt{1-x^2}}{x}\right| + C$

22. $\frac{1}{4}\left\{\text{Arctan}\frac{x}{2} - \frac{2x}{4+x^2}\right\} + C$ 24. $\frac{1}{3}\text{Arcsin}\frac{3(x-1)}{2} + C$

26. $\frac{1}{3}\ln\left|\sqrt{9x^2+36x+52} + 3(x+2)\right| + C$ 28. π

Section 26.9

2. $(1/49)(2 + 7x - 2\ln|2 + 7x|) + C$; 3

4. $\ln|x + \sqrt{x^2+6}| + C$; 35 6. $(1/60)(6x - 7)(7 + 4x)^{3/2} + C$; 11

8. $(1/6)\sin 3x + (1/14)\sin 7x + C$; 76

10. $(-1/4)\ln\left|\dfrac{4+\sqrt{16-x^2}}{x}\right| + C$; 26

12. $(e^{4x}/32)(8x^2 - 4x + 1) + C$; 54

14. $(-1/3)\ln\left|\dfrac{3+\sqrt{9-16x^2}}{4x}\right| + C$; 26

16. $\dfrac{1}{2(2+5x)} + \dfrac{1}{4}\ln\left|\dfrac{x}{2+5x}\right| + C$; 9

18. $(5/6)(x-1)\sqrt{2+4x} + C$; 13

20. $(e^{5x}/25)(5x - 1) + C$; 53

22. $(e^{5x}/29)(5\sin 5x + 2\cos 5x) + C$; 60

24. $(1/2)(\cos x^2 + x^2 \sin x^2) + C$; 81

26. $(1/5)\cos^4 x \sin x + (4/15)\cos^2 x \sin x + (8/15)\sin x + C$; 84

28. $\dfrac{x}{25\sqrt{25-4x^2}} + C$; 23

Section 26.10

2. 0.694 4. 0.406 6. 4.371 8. 1.397 10. 0.179 12. 225 m

14. 394 16. 3.749 18. 13.731 20. 0.747 22. 1.168 24. 0.768

26. a) 0.816 b) 0.843 28. a) 0.896 b) 0.861

30. a) 0.785 b) 0.785

Section 26.11

2. $20\pi/3$ 4. 9π 6. 1 8. $9\pi/2$ 10. π 12. 2π

14. $27\pi/2$ 16. $\dfrac{2^{2\pi}-1}{4\ln 2}$ 18. $4\pi + 12\sqrt{3}$ 20. 2π 22. $\dfrac{\pi-2}{8}$

24. $4\sqrt{3} - 4\pi/3$ 26. 4π 28. $(9/2)(3\pi - 8)$

Section 26.12

2. $\frac{1}{24}$ 4. 1 6. ∞ (diverges) 8. 1 10. ∞ (diverges) 12. $-\infty$ (diverges)

14. 1 16. 2 18. 1 20. $\frac{\pi}{2}$

CHAPTER 27

Section 27.1

2.

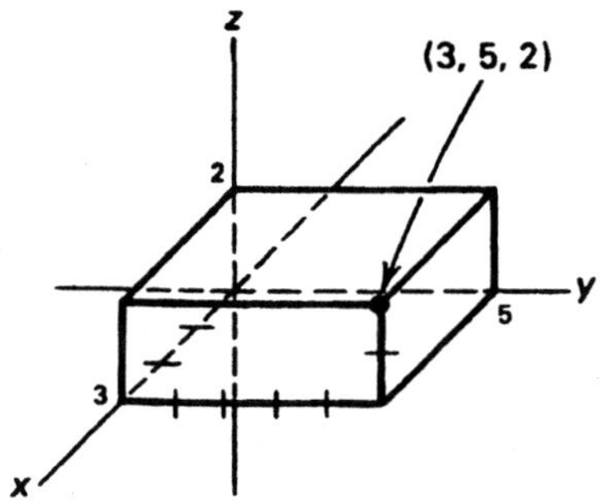

4.

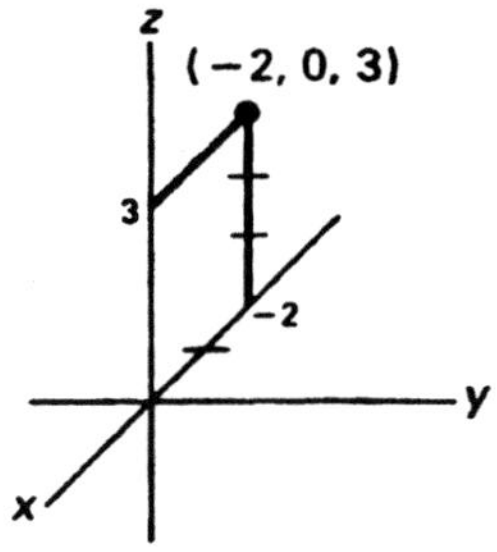

6.

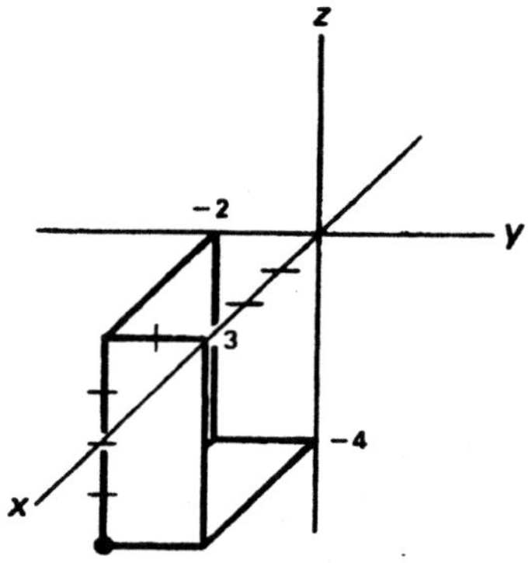

8.

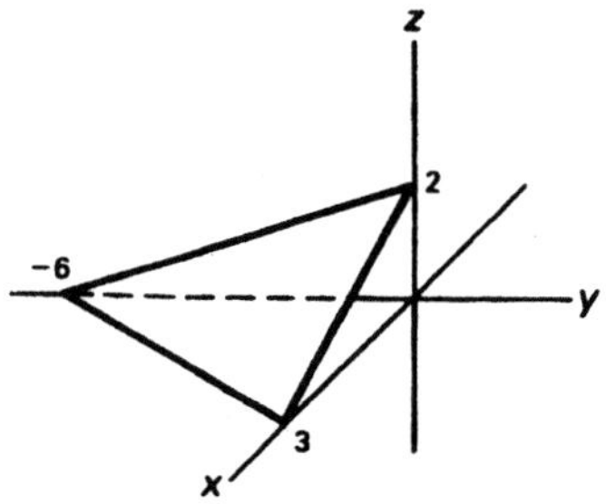

$6x - 3y + 9z = 18$

10.

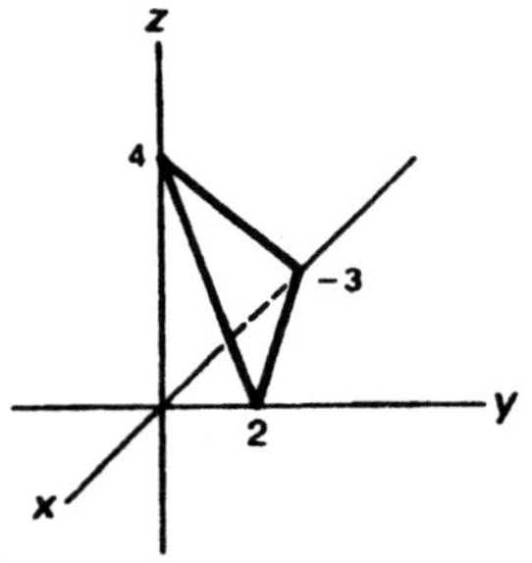

$-4x + 8y + 3z = 12$

12.

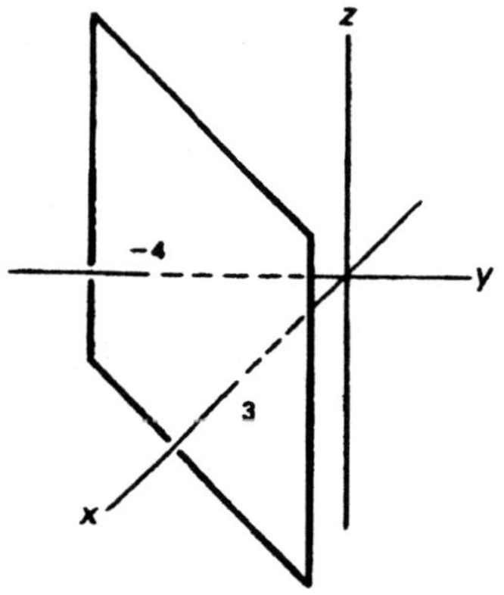

$4x - 3y = 12$

14.

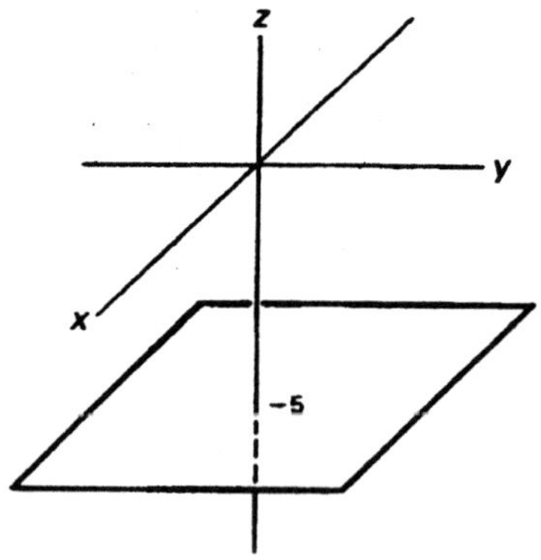

$z = -5$

16.

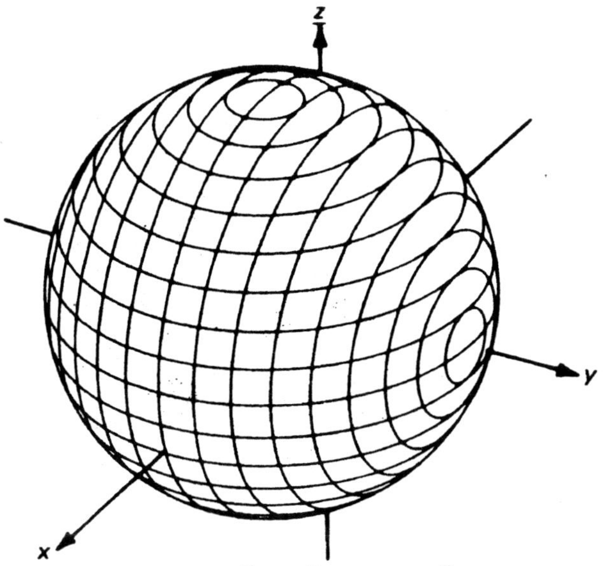

$(x-3)^2 + y^2 + (z-2)^2 = 36$

18.

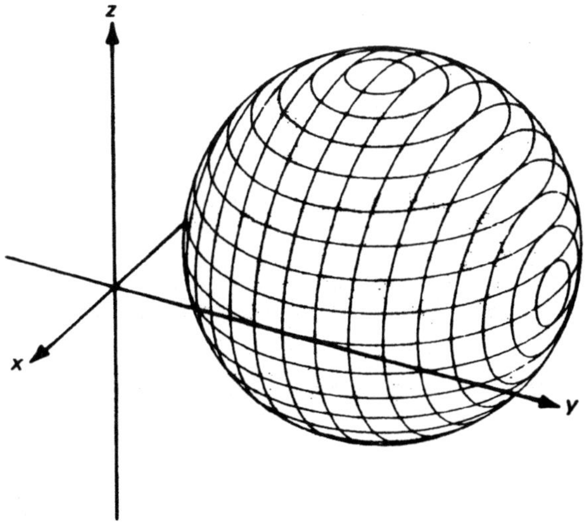

$x^2 + y^2 + z^2 + 6x - 6y + 9 = 0$

20.

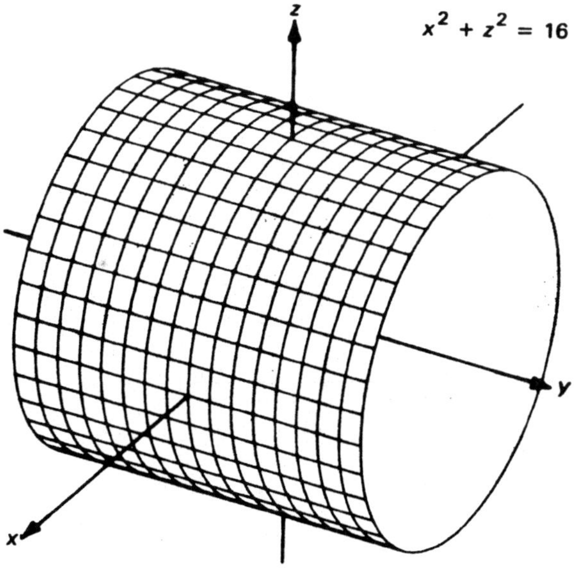

$x^2 + z^2 = 16$

22.

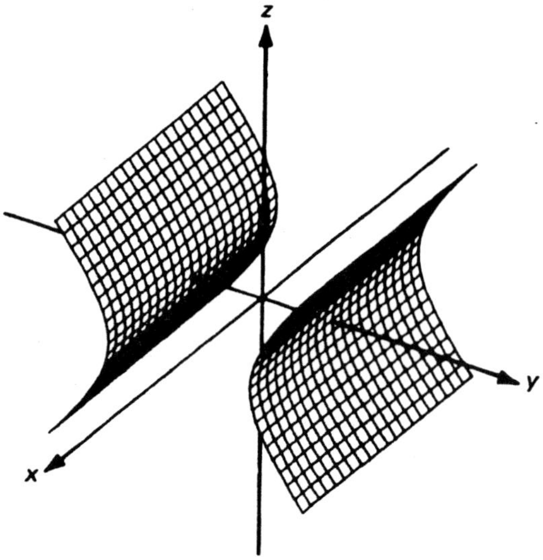

$y^2 - z^2 = 4$

24.

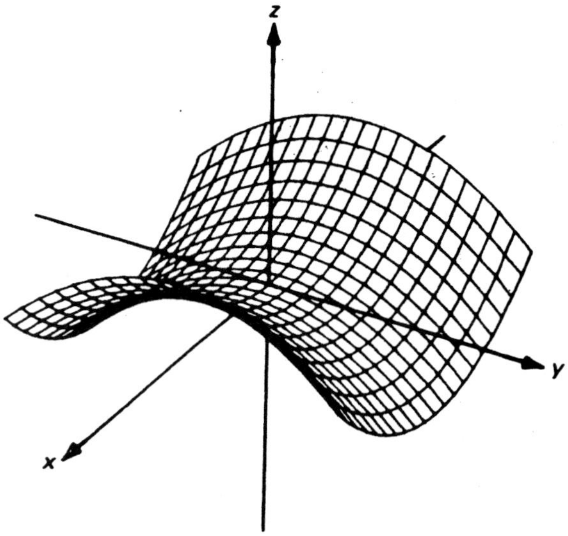

$x^2 - y^2 = 16z$

26.

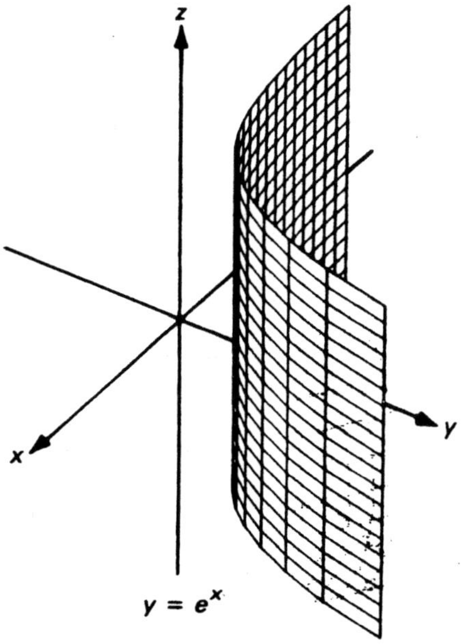

$y = e^x$

28.

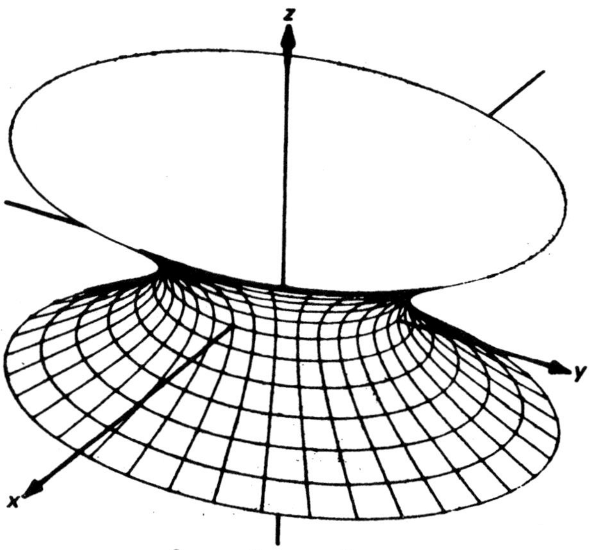

$16x^2 + 9y^2 - 36z^2 = 144$

30.

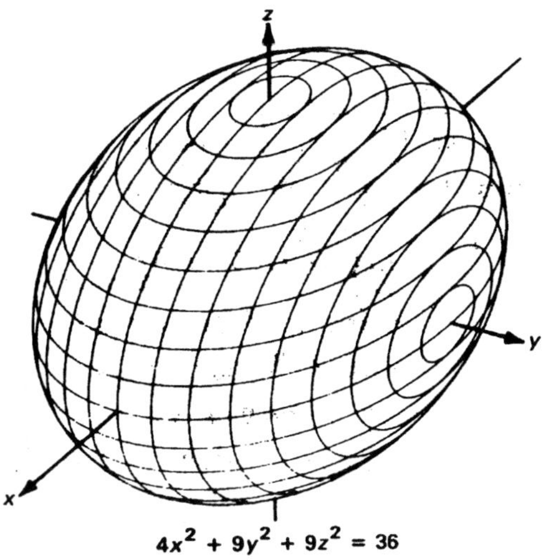

32.

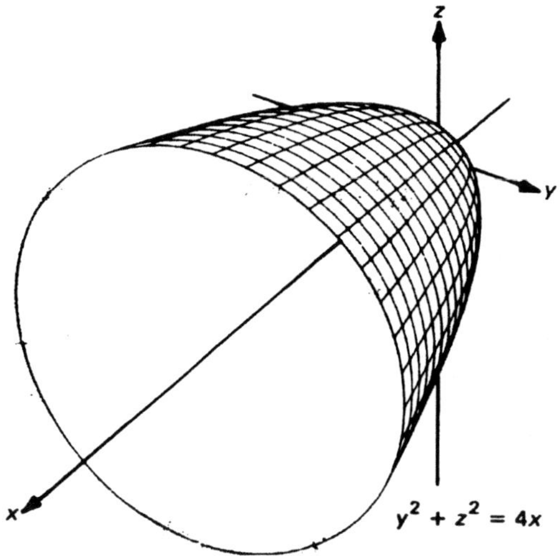

34.

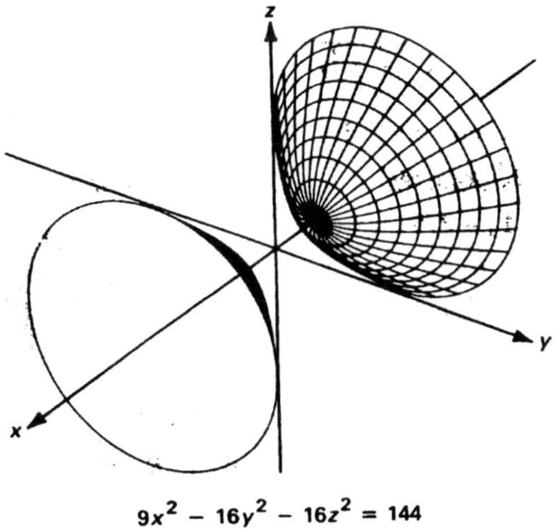

36.

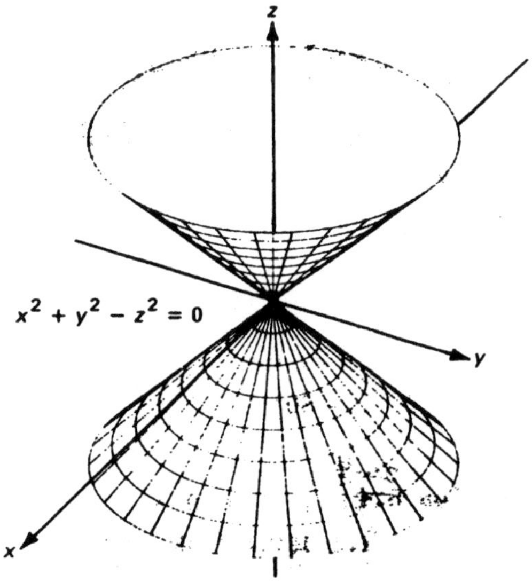

38. $(x-3)^2 + (y+2)^2 + (z-4)^2 = 4$ 40. $(x-4)^2 + (y-4)^2 + (z-4)^2 = 16$ 42. 9 44. $5\sqrt{5}$

Section 27.2

2. a) $16\pi x\sqrt{y}$ b) $4\pi x^2/\sqrt{y}$
4. a) $3y^5 - 8xy^2$ b) $15xy^4 - 8x^2y$
6. a) $\dfrac{x}{x^2+y^2}$ b) $\dfrac{y}{x^2+y^2}$
8. a) $2\pi xy/3$ b) $\pi x^2/3$
10. a) $ye^{xy} + e^{-x}$ b) xe^{xy}
12. a) $-6x(x^2-y^2)^{-4}$ b) $6y(x^2-y^2)^{-4}$
14. a) $\cos(x-y)$ b) $-\cos(x-y)$
16. a) $(1/y)\sec^2(x/y)$ b) $(-x/y^2)\sec^2(x/y)$

18. a) $y(\cos x)\, e^{\sin x}$ b) $e^{\sin x}$

20. a) $2x \cos 2y$ b) $-2x^2 \sin 2y$

22. a) $2xy \cos y$ b) $-x^2 y \sin y + x^2 \cos y$

24. $-V^2/R^2$

26. $(R_2)^2/(R_1 + R_2)^2$

28. $(X_L - X_C)/\sqrt{R^2 + (X_L - X_C)^2}$

30. R_2

32. $(E_1R_3^2 + E_1R_2R_3 - E_2R_3^2)/(R_1R_2 + R_1R_3 + R_2R_3)^2$

34. $\dfrac{tE}{R^2C^2}e^{-t/RC}$

36. $E - E\,e^{-t/RC} - \dfrac{Et}{RC}e^{-t/RC}$ or $E - \left\{E + \dfrac{Et}{RC}\right\}e^{-t/RC}$

38. $\dfrac{R}{R^2 + (X_L - X_C)^2}$

40. a) 4 b) 16

42. a) 2 b) −6

44. (8, −2, −28)

Section 27.3

2. $(2xy + 8xy^2 - 15x^2y^3)\,dx + (x^2 + 8x^2y - 15x^3y^2)dy$

4. $\dfrac{x-y}{x^2}dx + \dfrac{1}{x}dy$

6. $\dfrac{y}{1+x^2y^2}dx + \dfrac{x}{1+x^2y^2}dy$

8. $(y\,e^{xy} + y\cos xy)\,dx + (x\,e^{xy} + x\cos xy)\,dy$

10. 136 in^3

12. 8.37 ft

14. 0.5 A

16. min: (2, 4, −59)

18. max: (20, 20, 1200)

20. min: (−1, −3/2, −1/4)

22. max: (−1, −1, 5) saddle point: (0, 0, 4)

24. max: (−2, 0, 28) min: (4, 6, −188) saddle points: (4, 0 −80), (−2, 6, −80)

26. $\lambda = 2,\ w = 1,\ h = 3$

28. 10, 10, 10

Section 27.4

2. −3

4. 112/5

6. 5/4

8. 4

10. $(1/2)(e - 1)$

12. 2 ln 2 or ln 4

14. $2 + \pi^2/2$

16. 12

18. 16/3

20. 16/3

CHAPTER 28

Section 28.1

2. $0 - 3 - 8 - 15 - 24$
4. $-3 - 4 - 3 + 0 + 5 + 12$
6. $2 + \frac{2}{3} + \frac{1}{3} + \frac{1}{5} + \frac{2}{15} + \Lambda + \frac{4}{n(n+1)}$
8. $\frac{1}{2} - \frac{2}{3} + \frac{3}{4} - \frac{4}{5} + \Lambda$
10. $\sum_{n=1}^{22} (2n-1)$
12. $\sum_{n=2}^{n} (n-1)^3$
14. $\sum_{n=1}^{n} \frac{1}{2^n}$
16. $\sum_{n=0}^{n} \frac{n+1}{n+2}$
18. Diverges
20. Diverges
22. Converges
24. Diverges
26. Diverges
28. Diverges
30. Converges
32. Converges
34. Converges
36. Diverges
38. Converges
40. Diverges
42. Diverges
44. Converges

Section 28.2

2. Converges
4. Converges
6. Converges
8. Diverges
10. Diverges
12. Converges
14. Diverges
16. Diverges
18. Diverges
20. Diverges

Section 28.3

2. Converges absolutely
4. Diverges
6. Converges absolutely
8. Converges conditionally
10. Converges absolutely
12. Converges absolutely
14. Converges absolutely
16. Converges conditionally
18. Converges conditionally
20. Converges conditionally

Section 28.4

2. $-1 < x \le 1$
4. $-1 < x < 1$
6. $-1 \le x < 1$
8. $-4 < x < 4$
10. $-1/3 < x < 1/3$
12. $0 < x \le 2$
14. $-3/2 < x < -1/2$
16. $-5 \le x \le -1$
18. $-3 \le x \le 3$
20. $-\infty < x < \infty$
22. $-\infty < x < \infty$
24. $-\infty < x < \infty$

Section 28.5

2. $1-\frac{x^2}{2!}+\frac{x^4}{4!}-\Lambda$

4. $1+x+x^2+x^3+\cdots$

6. $1+3x+\frac{9x^2}{2!}+\frac{27x^3}{3!}+\Lambda$

8. $4x-\frac{64x^3}{3!}+\frac{1024x^5}{5!}-\Lambda$

10. $x^2-\frac{x^4}{3!}+\frac{x^6}{5!}-\Lambda$

12. $\frac{1}{3}-\frac{x}{54}+\frac{x^2}{324(2!)}-\frac{5x^3}{5832(3!)}+\Lambda$

14. $1-2x+\frac{6x^2}{2!}+\frac{24x^3}{3!}+\Lambda$

16. $1+\frac{3x}{2}+\frac{3x^2}{8}-\frac{x^3}{16}+\Lambda$

18. $16x^4-32x^3+24x^2-8x+1$ (finite sum)

Section 28.6

2. $1-\frac{x}{2!}+\frac{x^2}{4!}-\frac{x^3}{6!}+\frac{x^4}{8!}-\Lambda$

4. $x^2-\frac{x^6}{3!}+\frac{x^{10}}{5!}-\Lambda$

6. $1-4x+\frac{16x^2}{2!}+\frac{64x^3}{3!}+\Lambda$

8. $3x-\frac{9x^2}{2}+\frac{27x^3}{3}-\frac{81x^4}{4}+\Lambda$

10. $1-2x^2+\frac{4x^4}{2!}+\frac{8x^6}{3!}+\Lambda$

12. $x^3-\frac{x^5}{3!}+\frac{x^7}{5!}-\Lambda$

14. $e\left\{\frac{1}{x-1}+1+\frac{x-1}{2!}+\frac{(x-1)^2}{3!}+\Lambda\right\}$

16. −0.239814

18. 0.9045

20. 0.7040

22. $1+\frac{x^2}{2!}+\frac{x^4}{4!}+\Lambda$

24. 2499 V

Section 28.7

2. $\frac{\sqrt{2}}{2}\left\{1+(x-\pi/4)-\frac{(x-\pi/4)^2}{2!}-\frac{(x-\pi/4)^3}{3!}+\frac{(x-\pi/4)^4}{4!}+\Lambda\right\}$

4. $2+\frac{x-4}{4}-\frac{(x-4)^2}{32(2!)}+\frac{3(x-4)^3}{256(3!)}-\Lambda$

6. $1+2(x-\pi/4)+\frac{4(x-\pi/4)^2}{2!}+\frac{16(x-\pi/4)^3}{3!}-\Lambda$

8. $\frac{1}{e}\left\{1-(x-1)+\frac{(x-1)^2}{2!}-\frac{(x-1)^3}{3!}+\Lambda\right\}$

10. $(x-1)+\frac{(x-1)^2}{2!}-\frac{(x-1)^3}{3!}+\frac{2(x-1)^4}{4!}-\frac{6(x-1)^5}{5!}+\Lambda$

12. $\frac{1}{3}-\frac{2}{9}(x-1)+\frac{8(x-1)^2}{27(2!)}-\frac{16(x-1)^3}{27(3!)}+\Lambda$

14. $\frac{1}{2}-\frac{\sqrt{3}}{2}(x-\pi/3)-\frac{1}{2}\frac{(x-\pi/3)^2}{2!}+\frac{\sqrt{3}}{2}\frac{(x-\pi/3)^3}{3!}+\Lambda$

16. $e^3\left\{1-(x+3)+\frac{(x+3)^2}{2!}-\frac{(x+3)^3}{3!}+\Lambda\right\}$

Section 28.8

2. 0.7405
4. 0.03490
6. 0.40104
8. 0.9487
10. 1.97484
12. 2.4265
14. 3.03015

Section 28.9

2. $2\pi - 4\sin x - 2\sin 2x - (4/3)\sin 3x - \cdots$
   $f(x) = 2x,\ 0 \le x + 2\pi$

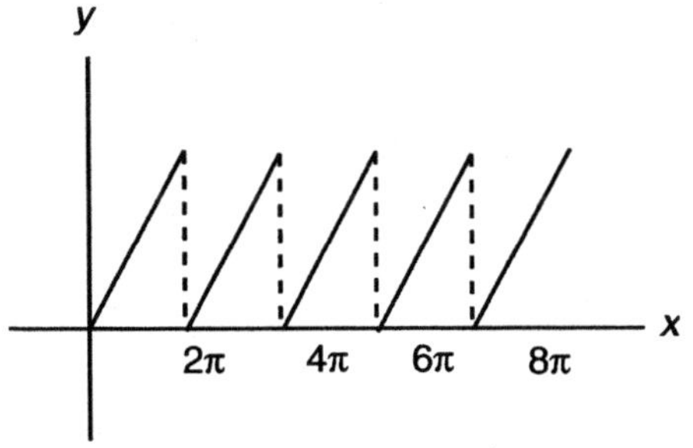

4. $4\sin x - 2\sin 2x + (4/3)\sin 3x - \cdots$
   $f(x) = 2x,\ -\pi \le x < \pi$

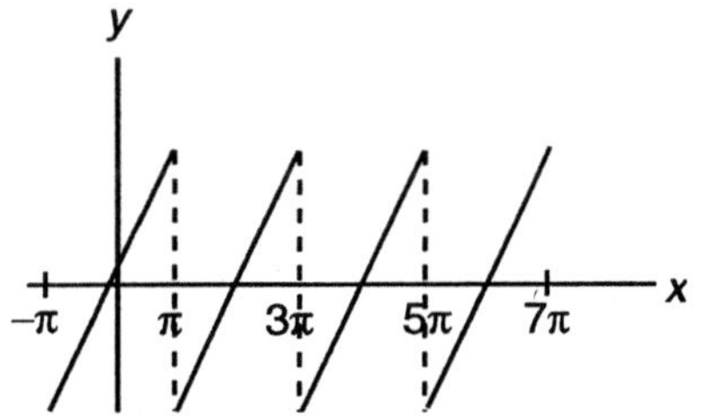

6. $\pi/2 + 2\sin x + (2/3)\sin 3x + (2/5)\sin 5x + \cdots$
   $f(x) = \begin{cases} \pi & 0 \le x < \pi \\ 0 & \pi \le x < 2\pi \end{cases}$

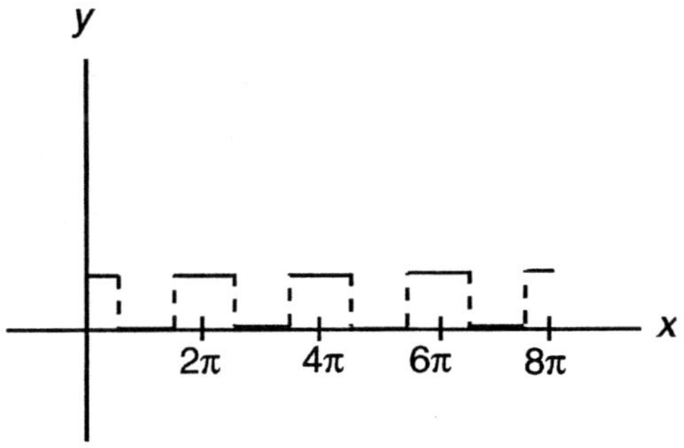

8. $\frac{3\pi}{4}-\frac{2}{\pi}\cos x-\frac{2}{9\pi}\cos 3x-\frac{2}{25\pi}\cos 5x-\Lambda\ -\sin x-\frac{1}{2}\sin 2x-\frac{1}{3}\sin 3x-(1/4)\sin 4x-\Lambda$

$f(x)=\begin{cases} x & 0\le x<\pi \\ \pi & \pi\le x<2\pi \end{cases}$

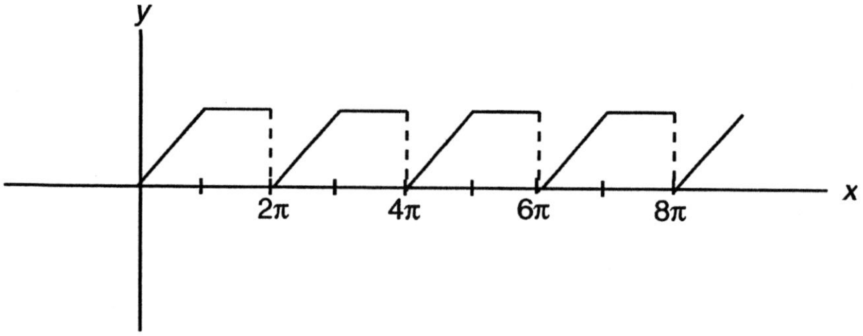

10. $\frac{-4}{\pi}\sin\frac{\pi x}{4}-\frac{4}{3\pi}\sin\frac{3\pi x}{4}-\frac{4}{5\pi}\sin\frac{5\pi x}{4}-\Lambda$

$f(x)=\begin{cases} 1 & -4\le x<0 \\ -1 & 0\le x<4 \end{cases}$

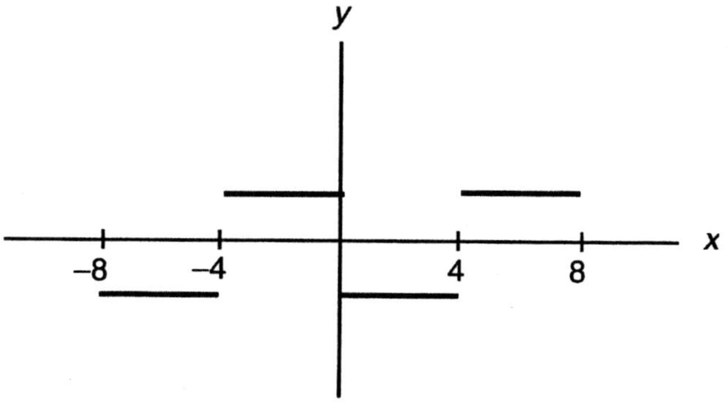

12. $\frac{3\pi}{4}+\frac{2}{\pi}\cos x+\frac{2}{9\pi}\cos 3x+\frac{2}{25\pi}\cos 5x+\Lambda\ -3\sin x\ -\frac{1}{2}\sin 2x-\sin 3x-\frac{1}{4}\sin 4x-\frac{3}{5}\sin 5x-\Lambda$

$f(x)=\begin{cases} 0 & 0\le x<\pi \\ x & \pi\le x<2\pi \end{cases}$

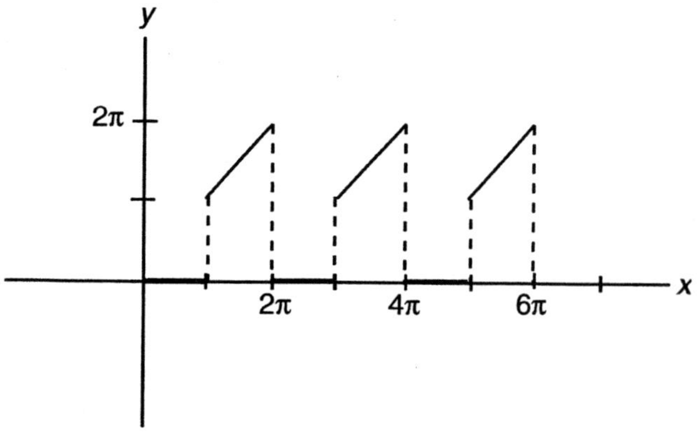

14. $\frac{-1}{4\pi}(e^{-4\pi}-1)-\frac{2(e^{-4\pi}-1)}{5\pi}\cos x-\frac{2(e^{-4\pi}-1)}{8\pi}\cos 2x-\frac{2(e^{-4\pi}-1)}{13\pi}\cos 3x-\Lambda-\frac{(e^{-4\pi}-1)}{5\pi}\sin x$

$-\frac{2(e^{-4\pi}-1)}{8\pi}\sin 2x-\frac{3(e^{-4\pi}-1)}{13\pi}\sin 3x-\Lambda$

$f(x)=e^{-2x},\ 0\le x\le 2\pi$

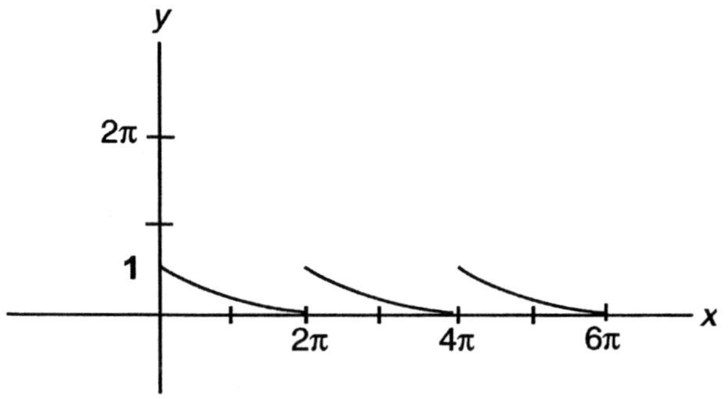

16. $\frac{2}{\pi}-\frac{4}{3\pi}\cos 2x-\frac{4}{15\pi}\cos 4x-\frac{4}{35\pi}\cos 6x-\Lambda$

$f(x)=\begin{cases}\sin x & 0\le x<\pi\\ -\sin x & \pi\le x<2\pi\end{cases}$

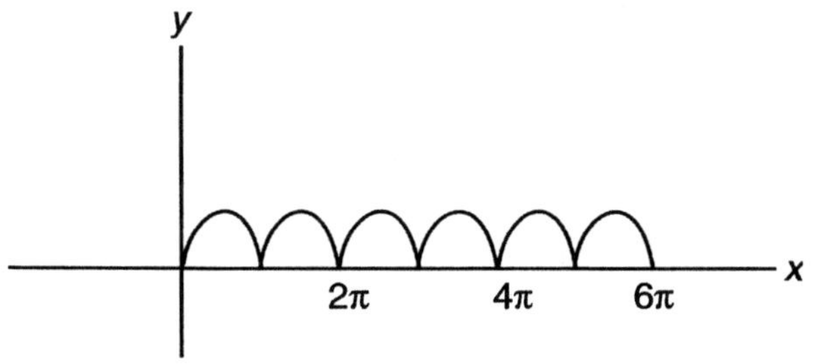

CHAPTER 29

Section 29.1

2. order 1; degree 2 4. order 1; degree 2 6. order 1; degree 1

8. order 2; degree 1

Section 29.2

2. $e^{x^3} = Cy$ 4. $\sin x = \cos y + C$ 6. $y^3/3 - 4y = x + C$

8. $y^2/2 = -\cos x + C$ 10. $-\ln(1 - y) = x + C$ 12. $y = C\sqrt{x^2 + 3}$

14. $y^3 = 3e^x + C$ 16. $x^2 = 6 \text{ Arcsin } y + C$

18. $y = (-1/2) \ln (1 + x^2) + C$ 20. $e^x + e^{-y} = C$

22. $e^{x^2} + e^{-2y} = C$ 24. $2 \text{ Arctan}(2x/3) = 3 \text{ Arcsin } 4y + C$

26. $(1 + x^4)^4(1 - y^2)^3 = C$ 28. $y^2 + 4e^x = 8$

30. $y = e^{[(x-1)/x]}$ 32. $y^3 = 6x - 12 \ln x + 21$

34. $e^{-2y} + x^2 - 4x + 2 = 0$ 36. $e^y = e^x + e - 1$

Section 29.3

2. $\ln\sqrt{x^2 + y^2} = x + C$ 4. $2(x^2 + y^2) = x^4 + C$

6. $3 \text{ Arctan } (y/x) = x^3 + C$ 8. $-5x + x^5y = Cy$

10. $\text{Arctan } (x/y) + \ln\sqrt{x^2 + y^2} = x + C$ 12. $3x = 3xy - 2y$

14. $4y = x^5 + 15x$

Section 29.4

2. $ye^{3x} = e^x + C$ 4. $y = x^3 - 5x + Cx^2$ 6. $y = -e^x + Ce^{x^3}$

8. $3y + 1 = Ce^{x^3}$ 10. $8x^2y = (x^2 + 4)^4 + C$

12. $2y \cos x + \cos^2 x = C$ 14. $3xy = 3x^3 \ln x - 5 + Cx^3$

16. $y = 2 + \dfrac{C}{(x + 1)^5}$ 18. $y(3x - 1)^{1/3} = 2(3x - 1)^{4/3} + C$

20. $(y-1)e^{\tan x} = C$

22. $2y = x^3 + 6x \ln x + 5x$

24. $e^{3x}(9y - 3x + 1) = 10$

26. $y = 1/2 + (3/2)e^{-8x^3/3}$ or $(2y-1)e^{8x^3/3} = 3$

28. $y(x-4)^{-3} = \ln(x-4) + 3$

30. $(y-1)e^{-\cos x} = 2$

Section 29.5

2. $v = t^3/3$; $v = 8/3$ m/s

4. $3y^4 = 4x^3 + 4$

6. $q = 3e^{-t/RC}$

8. $i = 0.4 - 0.4e^{-3000t}$

10. $Q = 2e^{-2.57\times10^{-6}t}$

12. $Q = 10e^{-0.00446t}$; 155 yr

14. 11.0 lb

16. 42.2° F

18. \$824.36

20. 21,600 h or 900 days

CHAPTER 30

Section 30.1

2. nonhomogeneous; 2
4. homogeneous; 5
6. nonhomogeneous; 3
8. homogeneous; 2
10. $y = k_1e^x + k_2e^{-5x}$
12. $y = k_1e^{3x} + k_2e^{-2x}$
14. $y = k_1e^{3x} + k_2e^{-3x}$
16. $y = k_1 + k_2e^{-2x}$
18. $y = k_1e^x + k_2e^{-5x/3}$
20. $y = k_1e^{5x/2} + k_2e^{-3x}$
22. $y = 1 - 2e^{-3x}$
24. $y = 5e^{2x} + 2e^{-5x}$
26. $y = 3 - e^{1-x}$

Section 30.2

2. $y = e^{-3x}(k_1 + k_2x)$
4. $y = e^x(k_1 \sin \sqrt{3}\, x + k_2 \cos \sqrt{3}\, x)$
6. $y = e^{4x}(k_1 + k_2x)$
8. $y = e^{-x}(k_1 + k_2x)$
10. $y = e^{-6x}(k_1 + k_2x)$
12. $y = k_1 \sin 4x + k_2 \cos 4x$
14. $y = k_1 \sin 4x/3 + k_2 \cos 4x/3$
16. $y = 3xe^{-5x}$
18. $y = \cos 4x - \sin 4x$
20. $y = 2e^{3x} \sin 4x$

Section 30.3

2. $y = k_1 \cos 2x + k_2 \sin 2x + (3/5)\ e^{-x}$
4. $y = e^{2x}(k_1 + k_2x) + e^x$
6. $y = k_1e^{3x} + k_2e^{-x} - x^2 + 4x/3 - 14/9$
8. $y = k_1 \sin x + k_2 \cos x + (1/2)e^x$
10. $y = k_1 \sin 2x + k_2 \cos 2x + (1/5\ e^x - x/2)$
12. $y = k_1e^x + k_2e^{3x} + 2 \cos x - 4 \sin x$
14. $y = k_1e^{3x} + k_2e^{-2x} + 1/36 - x/6 - (1/50) \sin x - (7/50) \cos x$
16. $y = k_1 \sin \sqrt{6}\ x + k_2 \cos \sqrt{6}\, x + 3x^2 - x - 1/2$
18. $y = k_1 \sin 2x + k_2 \cos 2x + (1/8)\ e^{-2x}$
20. $y = (1/2)(2 - e^{-x} - \sin x - \cos x)$
22. $y = (1/2)(2e^{2x} - e^{-2x} + 4x - 1)$

Section 30.4

2. $x = 0.5 \cos 8t$
4. $x = C_1 e^{-8t} + c_2 e^{-24t}$
6. 1640 lb
8. 2.11 ft
10. $i = e^{-1000t}(k_1 e^{995t} + k_2 e^{-995t})$
12. $i = 0.015e^{-200t} - 0.015e^{-1800t}$

Section 30.5

4. $\dfrac{1}{s-5}$
6. $\dfrac{49}{s(x^2+49)}$
8. $\dfrac{1}{(x-3)^2}$
10. $\dfrac{5}{(s-3)^2+25}$
12. $\dfrac{4s^2+9}{s^2(s^2+9)}$
14. $\dfrac{216}{s^2(s^2+36)}$
16. $120/s^6$
18. $\dfrac{s^2+8s-16}{(s^2+16)^2}$
20. $(s^2 - s + 2)\ \pounds\,(y)$
22. $(s^2 + 2)\ \pounds(y) + s - 2$
24. $(s^2 + 4)\ \pounds\,(y) - 2s + 3$
26. $(s^2 - 3s + 6)\ \pounds\,(y) - 7s + 24$
28. $(s^2 + 2s + 3)\ \pounds\,(y) + 4s + 3$
30. $(s^2 + 3s - 2)\ \pounds\,(y) + 5s + 18$
32. $\sin 4t$
34. $1 - e^{-2t}$
36. $t^3/6$
38. $t - e^{2t} \sin 3t$
40. $(1/3)\, e^{-2t} \sin 3t$
42. $8e^{-5t} + \cos 3t - (2/3) \sin 3t$
44. $6e^{3t} + 5e^{-2t}$
46. $4e^{-t} + 2te^{3t}$
48. $-2e^{-2t} + 2e^{-t/2} \cos \dfrac{\sqrt{11}}{2} t$

Section 30.6

2. $y = -e^{-3t}$
4. $y = 5e^{2t}$
6. $y = (-3/2)(1 - e^{-2t})$
8. $y = 2 \cos \sqrt{3}\, t + (5/\sqrt{3}) \sin \sqrt{3}\, t$
10. $y = (2/3)(1 - e^{-3t})$
12. $y = e^{3t}(1 - t)$
14. $y = (t^2/2)\, e^{-5t}$
16. $y = (1 - t + t^2)\, e^{3t}$
18. $y = (5/3)\, e^{2t} + (1/3)\, e^{-t}$
20. $i = 0.4 - 0.4e^{-3000t}$

APPENDIX B

Section B.1

2. centi 4. deci 6. deka 8. micro 10. k 12. d 14. da 16. μ

18. 63 dag 20. 42 cm 22. 25 mg 24. 17 dL 26. 182 litres

28. 147 kilograms 30. 23 decilitres 32. 17 milligrams 34. 250 micrograms

36. kilogram 38. ampere 40. watt

Section B.2

2. 1 metre 4. 1 centimetre 6. 1 kilometre 8. mm 10. km 12. cm

14. m 16. m 18. cm 20. mm 22. m 24. mm 26. cm

28. cm 30. cm 32. mm, mm 34. 0.001 36. 0.01 38. 10 40. 10

42. 0.01 44. 0.1 46. 23,000 cm 48. 57.6 cm 50. 0.25 dm 52. 8.5 km

54. 8500 μm 56. Answers vary 58. Answers vary

Section B.3

2. 1 gram 4. 1 centigram 6. 1 kilogram 8. mg 10. g 12. mg

14. kg 16. kg 18. kg 20. metric ton 22. g 24. kg 26. g

28. g 30. g 32. g 34. mg 36. 0.001 38. 0.01 40. 10

42. 10 44. 0.01 46. 0.1 48. 1000 50. 0.565 kg 52. 275,000 g

54. 48,500 dg 56. 5000 kg 58. 5.7×10^{-5}g 60. 40 metric tons

Section B.4

2. 1 kilolitre 4. 1 cubic metre 6. 1 square decimetre 8. L 10. cm^2 12. L

14. cm^2 16. *ha* 18. L 20. L 22 *ha* 24. mL 26. cm^2 28. m^3

30. cm^2 32. m^3 34. *ha* 36. mL 38. 1000 40. 0.001 42. 10 44. 0.001

46. 1000 48. 1 50. 1 52. 10^6 54. 10^{-6} 56. 10,000 58. 0.01 60. 800 mL

62. 0.008 L 64. 6×10^6 cm^3 66. 0.53 cm^3 68. 0.145 L 70. 1.6×10^{-4} L

72. 0.35 m^3 74. 1.25×10^6 m^2 76. 15,000 mm^2 78. 5×10^3 m^2 80. 1000 kg

Section B.5

2. b 4. c 6. b 8. d 10. c 12. c 14. a 16. 15

18. 707 20. –15 22. 10.4 24. 323 26. 27°C 28. –321°F 30. 1204

32. 941 34. 1175 36. 590 38. ampere; A 40. 1 watt 42. 1 millisecond

44. 1 μs 46. 1 mW 48. 3.6 μs 50. 150 mA 52. 12 A 54. 75 ps 56. 10^6

58. 10^{-3} 60. 10^{-6} 62. 10^3 64. 10^3 66. 2.5 kW 68. 2.45×10^{-4}s

70. 85 min 72. 1944 h 74. 6×10^4W 76. 3.6×10^{15} ps

Section B.6

2. (a) 20,600 oz (b) 586 kg (c) 586,000 g 4. (a) 250,000 g (b) 550 lb

(c) 2.50×10^8 mg 6. (a) 520 qt (b) 493 L (c) 0.493 kL

8. (a) 8.60×10^8 cm^2 (b) 0.0860 km^2 (c) 103,000 yd^2 10. (a) 25,000 L

(b) 883 ft^3 (c) 1.53×10^6 in^3 (d) 25 m^3 12. (a) 371 acres (b) 1.50×10^6 m^2

(c) 1.50 km^2 (d) 1.62×10^7 ft^2 14. 88.6 km/h 16. 1.8°C/min 18. 27 in./h

20. 1.98×10^5 km/h^2 22. 5.75 g/cm^3 24. 2.50×10^{-5} g/mm^2 26. 3.35 hp 28. 13.7 slugs

30. 1.38 in. 32. 11.4 cm 34. 605 lb 36. 1850 mg 38. Answers vary

40. 78.5 cm^3

Chapter Tests

Chapter 1 Test

1.1

Perform the indicated operations where possible and simplify.

1. $(-5)+(-3)$
2. $(-8)(9)$
3. $(-8)-(-5)$
4. $\left(5\frac{3}{5}\right)\div\left(-1\frac{7}{15}\right)$

1.2

5. $\dfrac{-8\cdot 0}{-4+(-4)}$
6. $24\div 8+5-6(2)$
7. $3\cdot 24\div 12\cdot 2\div 4$
8. $\dfrac{27\div 3\cdot 5\div 15}{3-(-5)}$

1.3

9. $\dfrac{10^4}{3-3}$
10. $\dfrac{10^{-3}\cdot 10^{4}\cdot 10^{-7}}{10^{-5}\cdot 10^{2}}$
11. Write in scientific notation: 3,870,000
12. Write in decimal form: 2.65×10^{-5}
13. Find the value and round the result to three significant digits written in scientific notation.

$$\frac{(-8.41\times 10^{-8})(2.52\times 10^{-7})}{(-3.27\times 10^{-5})}$$

1.4

14. Find the measurement that is (a) the most accurate and (b) the most precise.

    2384.0 km; 50.00 km; 0.00007 km; $28\bar{0}$ km

15. Find the measurement that is (a) the least accurate and (b) the least precise.

    0.000003 A; 0.0408 A; 0.78940 A; 0.057 A

1.5

16. Use the rules for addition of measurements to find the sum.

    8.523 cm + 18.24 cm + 8.096 cm + 0.0952 cm

17. Use the rules for multiplication of measurements to evaluate the following.

    (0.672 m)(0.0845 m)(0.0093 m)

1.6

Perform the indicated operations and simplify.

18. $(-4x + 8x^2 + 7) - (5x - 9 - 3x^2)$

19. $(3x^4 - 5x + 4x^3) + (3x^2 - 9x^4 - 6x)$

Evaluate each expression when $x = -5, y = -4, w = 3$.

20. $\dfrac{2x - 3y}{x - w}$

21. $\dfrac{(xw - 2y)^2}{w - y}$

1.7

Perform the indicated operations and simplify.

22. $x^3 \cdot x^4 \cdot x^7$

23. $\dfrac{w^{14}}{w^8}$

24. $(-t^3)^4$

25. $(2x^2y^3)^5$

Evaluate and round to three significant digits.

26. $\sqrt{100-49}$

27. $\sqrt{6^2+9^2}$

28. $\sqrt{(7.5\times 10^2)^3+(6.2\times 10^2)^4}$

1.8

Find each product.

29. $(3x-4)(3x+4)$

30. $(5x^2-3)^2$

1.9

Find each quotient.

31. $\dfrac{35x^2y^5}{49x^8y^2}$

32. $\dfrac{45x^4-27x^3-18x^2+33x}{3x}$

33. $\dfrac{2x^3-11x^2+22x-21}{x-3}$

1.10

Solve each equation.

34. $3(2x-5)+4x=4x+9$

35. $3.29x+27.6=0.622$

1.11

Solve for the indicated letter.

36. $x=\dfrac{Q}{P}$ for Q

37. $aF=hx-w$ for x

1.12

38. Given $\dfrac{1}{f}=\dfrac{1}{S_0}+\dfrac{1}{S_i}$ where $f = 36$ cm and $S_0 = 5\bar{0}$ cm, find S_i.

39. Given $\dfrac{1}{R}=\dfrac{1}{R_1}+\dfrac{1}{R_2}$ where $R = 50.0\ \Omega$ and $R_1 = 24\bar{0}\ \Omega$, find R_2.

1.13

40. Lee has two containers of solder (a mixture of tin and lead) available. One is 25% tin and the other is 60% tin. How much of each must be used to make a 15-lb mixture that is 45% tin? Round your answer to three significant digits.

41. A bar of metal contains 10% gold; another bar contains 18% gold. How many kilograms of each must be used to make a 20-kg bar containing 15% gold? Round your answer to three significant digits.

1.14

Solve each proportion. Round each answer to three significant digits.

42. $\frac{x}{4.7-x}=\frac{3}{4}$

43. $\frac{x}{8.5-x}=\frac{5}{8}$

44. On an assembly line $70{,}20\overline{0}$ bolts are to be divided into two groups in the ratio of 4 to 5. How many bolts will be in each group?

45. A fuel pump delivers 56.00 mL of fuel in 588 strokes. How many strokes are needed to use 116.0 mL of fuel?

1.15

46. y varies directly with w; $y = 209$ when $w = 55$. Find k, the constant of variation. Find y when $w = 220$.

47. m varies directly with the square of x and inversely with y; $m = 20.8$ when $x = 12.0$ and $y = 45.0$. Find k, the constant of variation. Find m when $x = 24.0$ and $y = 26.0$.

Chapter 2 Test

2.1

For problems 1-3 give the mathematical name associated with the description.

1. Lines that do not intersect, even when extended.

2. Two angles whose measures have a sum of 90°.

3. Opposite angles formed by two intersecting lines.

For problems 4-5, parallel lines *l* and *m* are cut by a transversal, *t*. Given angle a=105°, find the following

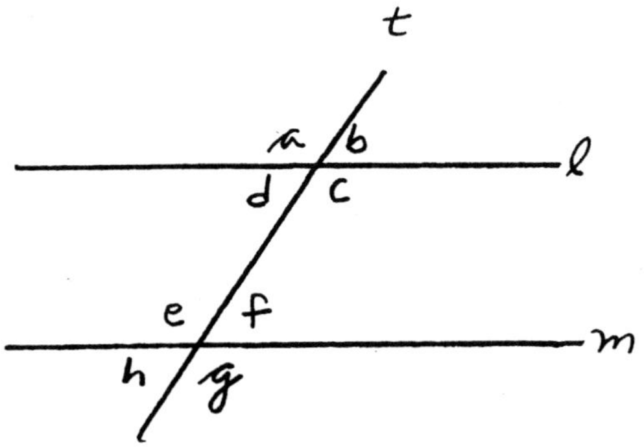

4. angle b

5. angle g

2.2

For problems 6-8 give the mathematical name associated with the description.

6. A triangle in which exactly two sides are equal.

7. A perpendicular line segment from any vertex of a triangle to the opposite side.

8. The sum of the lengths of the sides of a polygon.

9. The legs of a right triangle measure 4.92 cm and 6.35 cm. Find the length of the hypotenuse.

10. Find the area of a triangle with a base of 9.27ft and an altitude of 5.93 ft.

11. Two of the three angles of a triangle have measurements of 36.7° and 85.9°. What is the measure of the third?

12. A 17.2 ft tree is transplanted. A guy wire is attached 3.5 ft from the top and anchored

to the ground 11.8 ft from the base of the tree. How long is the guy wire?

2.3

13. A lot in the shape of a parallelogram must be fenced and fertilized. The sides of the lot are 368 ft and 292 ft with an altitude of 263 ft to the base of 368 ft.

a. If fencing is installed at $8.75 per running foot, find the cost of fencing the lot.

b. If one bag of fertilizer covers 2570 ft^2 and costs $17.95, find the cost to fertilize the grass in the lot.

14. The area of a trapezoid is 3112.5 m^2, its height is 37.5 m, and one of its bases is 69.2 m long. Find the length of the other base.

2.4

15. Find the circumference and the area of a circle if the length of its radius is 36.5 m.

16. Find the diameter of a circle where the area is 17.95 cm^2.

For problems 17-19, use figure where $\angle BAC = 25.0°$.

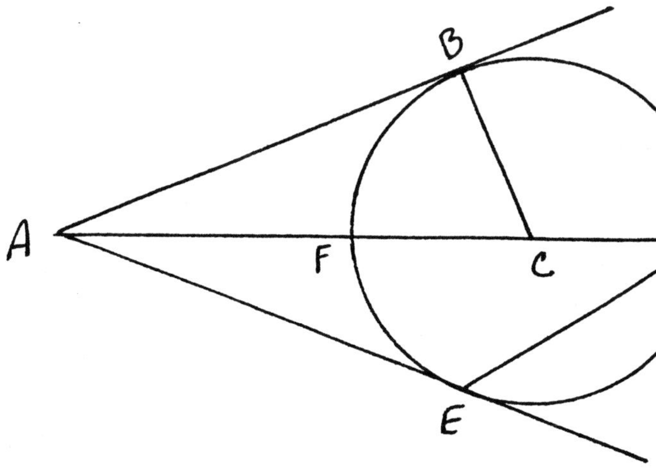

17. Find BF

18. Find EF

19. Find $\angle ADE$

2.5

20. Find a) the volume and b) the lateral surface area of a right circular cone, given r = 2.60 m and h = 5.80 m.

21. Find a) the volume and b) the total surface area of a sphere whose diameter si 96.0 ft.

22. Find a) the volume and b) the total surface area of a right circular cylinder 24.6 cm high and 14.4 cm in diameter.

Chapter 3 Test

3.1

In problems 1-6, round your answer to three significant digits.

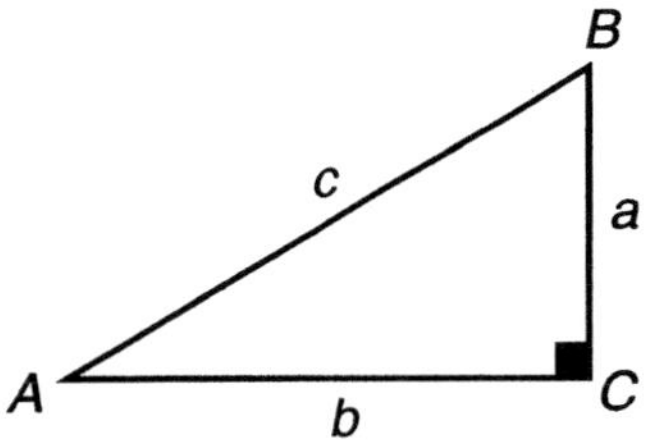

1. Given $a = 23.8$ cm, $c = 46.2$ cm, find b.

2. Given $a = 4.92$ mm, $b = 7.35$ mm, find c.

3. Given $a = 17.6$ inches, $b = 20.4$ inches, $c = 26.9$ inches, find sin A.

4. Given $a = 3.27$ cm, $b = 7.84$ cm, $c = 8.49$ cm, find cos A.

5. Given $a = 384$ m, $b = 257$ m, $c = 462$ m, find tan A.

6. Given $a = 49.3$ in., $b = 35.9$ in., $c = 61.0$ in., find csc B.

3.2

Find the value of each rounded to four significant digits.

7. sin 57.3E   8. tan 38E 25'   9. sec 67.8E

10. cot 70E 47' 18"   11. cos 18E 39' 40"   12. csc 59.6E

Find each angle rounded to the nearest hundredth of a degree.

13. $\sin \theta = 0.9790$ 14. $\cos \theta = 0.7363$

15. $\cot \theta = 1.8871$ 16. $\csc \theta = 1.2564$

3.3

Solve each right triangle. Round each side to three significant digits and each angle to the nearest tenth of a degree.

17. $a = 39.6$ ft, $c = 52.5$ ft   18. $b = 6.43$ cm, $A = 21.7$E

3.4

19. A tree casts a shadow 30.9 ft long. If the angle of elevation from the tip of the shadow to the top of the tree is 61.4E, how high is the tree?

20. An engineer has decided that for safety, a road should have an angle of inclination of at most 3.7E. For a horizontal section measuring 3540 km, what is the highest value the vertical measurement may be?

21. At an altitude of $36\bar{0}0$ ft, a pilot enters her final landing approach by setting the angle of depression to 24.8E. How far will she travel through the air before touching down?

22. In an ac circuit the reactance is 95.3 Ω and the resistance is 126 Ω. Find the impedance and the phase angle to the nearest tenth of a degree.

Chapter 4 Test

4.1

Determine whether or not each relation is a function. Write its domain and range.

1. $\{(-3, 4), (-3, 6), (-3, 8), (-3, 10)\}$

2. $y = 4x^2 - 1$

3. $y = 5 - \sqrt{6x - 12}$

4. $x = 7 - y^2$

5. Given $f(x) = \dfrac{3x - x^2}{x}$, find

   a) f(2) b) f(−4) c) f(3a)

4.2

Graph each equation.

6. $3x - 2y = 12$

7. $y = x^3 - 5x^2 + 12$

4.3

8. Find the slope of the line passing through the given points (−7, −4) and (−1, −9).

9. Find the slope of the line perpendicular to $y = \dfrac{3}{5}x + 2$.

10. Find the equation of the line that passes through (3, −7) with slope $-\dfrac{1}{2}$.

4.4

11. Find the equation of a line that is perpendicular to the x-axis and passes through the point (5, −3).

12. Find the equation of a line that passes through the point (3, −8) and is parallel to $2x - y = 6$.

13. Determine whether the given pair of equations represents lines which are parallel, perpendicular, or neither.

$2x - 3y - 3 = 0$ and $3x + 2y - 8 = 0$

14. Find the distance between the points (7, –3) and (–5, –8).

15. Find the coordinates of the point midway between (–3, –5) and (7, –11).

16. Determine whether or not triangle ABC is a right triangle.
A (–1.6, 4.2), B (8,9), C (5, –9).

Chapter 5 Test

5.1

Find each product.

1. $(3x-5)(2x+7)$ 2. $4x^2(x-6)$ 3. $(2x+3)^2$

5.2

Factor each expression completely.

4. $6x-15$ 5. $w^2+14w+49$ 6. $x^2-4x-21$

7. $2x^2+15x+18$ 8. $20x^2+26x-60$ 9. $4x^2-25$

5.3

Factor each expression completely.

10. $a(3x+y)+2m(3x+y)$ 11. $8x^3-125$

4.4

Reduce each fraction to lowest terms.

12. $\dfrac{45x^2y}{54xy^2}$ 13. $\dfrac{8x+12}{10x+15}$

14. $\dfrac{6x^2-37x-35}{x^2-49}$ 15. $\dfrac{6+13x-5x^2}{x^2+3x-18}$

5.5

Perform the indicated operations and simplify.

16. $\dfrac{(3x+4)(x-2)}{x+7}\cdot\dfrac{x^2+8x+7}{3x^2-11x-20}$ 17. $\dfrac{2x^2+7x-4}{2x^2-9x+4}\div\dfrac{x+4}{x^3-64}$

5.6

18. Find the lowest common denominator for the set of denominators.

$4x^2 + 12x;\ x^2 + 6x + 9;\ 2x^2$

Perform the indicated operations and simplify.

19. $\dfrac{6x}{3x+4} + \dfrac{8}{3x+4}$

20. $\dfrac{1}{x+5} - \dfrac{3}{2x+7} + \dfrac{3}{(2x+7)(x+5)}$

21. $\dfrac{20x}{9x^2-4} + \dfrac{x-4}{3x-2}$

5.7

Perform the indicated operations and simplify.

22. $\dfrac{\dfrac{3}{x} - \dfrac{3}{y}}{\dfrac{3}{x} + \dfrac{3}{y}}$

23. $\dfrac{\dfrac{6x^2-13xy-5y^2}{3x+y}}{\dfrac{2x^2+3xy-20y^2}{2x+7}}$

5.8

Solve and check.

24. $\dfrac{x-8}{x+2} = \dfrac{x+5}{x-3}$

25. $\dfrac{3}{x-6} + \dfrac{1}{x+4} = \dfrac{x-12}{(x-6)(x+4)}$

Solve for the indicated letter.

26. $F = \dfrac{mv^2}{r}$ for m

27. $x = \dfrac{3}{c} - \dfrac{3}{t}$ for t

Chapter 6 Test

6.1

Solve each system of equations.

1. $x + y = 2$
   $2x - y = 7$

2. $x + 2y = 6$
   $3x + y = -7$

3. $x = 2.5y + 3.5$
   $2.4x - 6y = 9$

4. The sum of two capacitors is 70 μF. The difference between them is 20 μF. What is the size of each capacitor?

5. A hospital worker has a 5% solution and a 12% solution of a disinfectant. How much of each must she mix to get $150\bar{0}$ mL of a 10% solution? Round your answers to the nearest whole mL.

6. A rushed subway traveler got on an up-escalator and also walked up. After 45 seconds he realized that he had left his briefcase at the bottom. Since he would not have access to a down-escalator, he turned and walked down the up-escalator, reaching the bottom in two minutes. If he had traveled up 288 feet, find the rate of the escalator and the unaided rate of the man in feet per second.

6.2

7. Solve the system.

$$\frac{2}{x} + \frac{1}{y} = 2$$

$$\frac{3}{x} - \frac{2}{y} = 31$$

6.3

Solve each system of equations.

8.
$$\begin{aligned} x + y + 2z &= 10 \\ 3x - y + x &= -9 \\ 4x + 3y - 5z &= -5 \end{aligned}$$

9.
$$\begin{aligned} x + y + z &= 4.7 \\ 5x + 2y - z &= -2 \\ -5x + 4y + 3z &= -17 \end{aligned}$$

10. A manufacturer is making trail mix using pecans which cost \$6.90 per pound, raisins

which cost $3.50 per pound and coated chocolate candy which costs $4.50 per pound. He plans to make 100 pounds of mix worth $5.11 per pound. If he uses 10 more pounds of raisins than he does of candy, how many pounds of each should he use?

6.4

Evaluate each determinant.

11. $\begin{vmatrix} 8 & 3 \\ 4 & 5 \end{vmatrix}$

12. $\begin{vmatrix} m & 6 \\ 4 & 5 \end{vmatrix}$

13. $\begin{vmatrix} 2 & 3 & -7 \\ -5 & 0 & -1 \\ 4 & 1 & 6 \end{vmatrix}$

14. $\begin{vmatrix} 4 & 0 & -5 & 3 \\ -3 & 5 & 7 & -2 \\ 0 & 6 & 8 & 9 \\ 1 & 2 & 0 & 1 \end{vmatrix}$

6.5

Problems 15-18 are true or false.

15. In a determinant, A, if every element of a row is multiplied by 6, then the value of A is multiplied by 6.

16. If two columns of a determinant are identical, then the value of the determinant is zero.

17. Two adjacent rows of a determinant may be interchanged without changing the value of the determinant.

6.6

18. When using Cramer=s Rule to solve a system of linear equations, it cannot be determined if the system is inconsistent.

19. Using 2×2 determinants, solve the expression for the x-coordinate of the solution to the system:

$$f_1x + g_1y = 7$$
$$f_2x + g_2y = 10$$

20. Use Cramer=s Rule to solve for the y-coordinate of the solution to the system:

$$2x + 3y - z = -5$$

$$4x - 3y + 2z = -4$$
$$x + 6y + 5z = 17$$

6.7

Find the partial fractions of each expression.

21. $\dfrac{4x+36}{(x-3)(x+5)}$

22. $\dfrac{x^3+33x^2-12x+108}{x^4-16}$

23. $\dfrac{8x+10}{(2x+5)^2}$

Chapter 7 Test

7.1

Solve each equation.

1. $2x^2 - x = 21$

2. $4x^2 - 27 = 0$

3. $\dfrac{4x+49}{16x} = x + \dfrac{1}{4}$

7.2

4. Find the value of k which will make each expression a perfect binomial square.

   a) $x^2 + 10x + k$

   b) $x^2 - \dfrac{x}{3} + k$

Solve each equation by the method of completing the square.

5. $x^2 + 3x - 10 = 0$

6. $6x^2 + x - 15 = 0$

7.3

Use the quadratic formula to solve each equation.

7. $2x^2 + 13x + 18 = 0$

8. $3x^2 + 5x = 28$

9. $8x^2 + 6x = 35$

10. $6x^2 - x - 15 = 0$

7.4

Write an equation corresponding to each problem, then solve. Express your answers using three significant digits.

11. Design a rectangular glass plate whose area is 864 in^2 and whose length is 6 in. less than three times its width.

12. A variable voltage in a given electrical circuit is given by the formula $V = t^2 - 13t + 50$. At what value of t (in seconds) is the voltage V equal to 20.0 volts? to 35 volts?

13. The perimeter of a rectangular garden is 25.6 m and its area is 36.12 m^2. Find the

length and width of the garden.

14. A rectangular container is to be made from a 46.2 cm × 70.2 cm rectangular sheet of metal by cutting a square out of each corner and folding up the sides. Find the dimensions of each square if the area of the bottom is to be 2065 cm^2.

Chapter 8 Test

8.1

Simplify and write each result using positive exponents.

1. $x^8 \cdot x^{10}$

2. $x^7 \cdot x^{-9}$

3. $\dfrac{x^8}{x^5}$

4. $(2x)^3$

5. $(x^{-2})^{-4}$

6. $(3x^2w^3)^4$

7. $\dfrac{x^{-5}y^4}{x^4y^{-3}}$

8. $\left(\dfrac{8x^{-5}w^{-2}}{4x^{-7}w^2}\right)^{-3}$

8.2

Simplify and write each result using positive exponents.

9. $81^{1/2}$

10. $(-27)^{-1/3}$

11. $125^{2/3}$

12. $2x^{-1/2}\left(5x^{3/2} - 3x^{-3/2}\right)$

8.3

Simplify each radical expression. Assume all variables represent positive real numbers.

13. $\sqrt{48}$

14. $\sqrt{162x^2}$

15. $\sqrt{\dfrac{3}{32}}$

16. $\dfrac{8}{\sqrt{20y^3}}$

17. $\sqrt[3]{135x^4w^5}$

18. $\sqrt[3]{\dfrac{5}{192x^2y}}$

8.4

Perform the indicated operations and simplify when possible. Assume all variables represent positive real numbers.

19. $\sqrt{48}+5\sqrt{75}-2\sqrt{108}$

20. $3\sqrt{80x^3}-2\sqrt{500x^3}+6x\sqrt{45x}$

21. $\dfrac{8}{\sqrt{8x^3}}+\dfrac{15}{\sqrt{50x^3}}-\dfrac{20}{x\sqrt{98x}}$

8.5

Perform the indicated operations and simplify when possible. Assume all variables represent positive real numbers.

22. $\sqrt{18}\sqrt{175}$

23. $\dfrac{\sqrt{80}}{\sqrt{5}}$

24. $\dfrac{\sqrt{24}}{\sqrt[4]{4}}$

25. $\left(8-3\sqrt{7}\right)\left(2+\sqrt{7}\right)$

26. $\left(3\sqrt{5}-\sqrt{6}\right)^2$

8.6

Solve and check.

27. $6\sqrt{x-5}=45$

28. $\sqrt{35-2x}=x$

29. $\sqrt{6x^2-13x}=2\sqrt{7}$

30. $\sqrt{x^2+3x+11}=x+3$

8.7

Solve each equation.

31. $2x^4-23x^2+45=0$

32. $x^{-4}-29x^{-2}+100=0$

33. $(x-3)^4-20(x-3)^2+64=0$

Chapter 9 Test

9.1

1. Graph $y = 7^x$. Give four ordered pairs which are on the graph.

2. Find $4^{-\pi}$ rounded to three significant digits.

3. The formula for compound interest is $A = P\left(1+\frac{r}{x}\right)^{xn}$. Lee opened a retirement account with an initial investment of \$10,000 at an annual interest rate of 8.5%. If the interest is compounded daily, how much will be in the account at the end of 25 years?

9.2

4. Write $3^w = m$ in logarithmic form.

5. Write $\log_c 4 = f$ in exponential form.

6. Graph $y = \log_3 x$. Give four ordered pairs which are on the graph.

For problems 7-10, match each graph with the appropriate equation.

A.
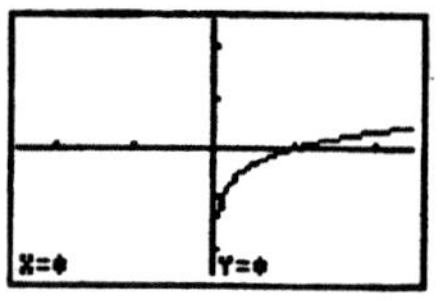

B.

C.
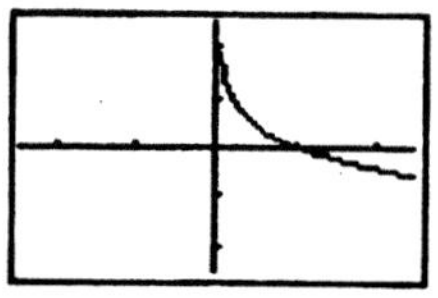

D.
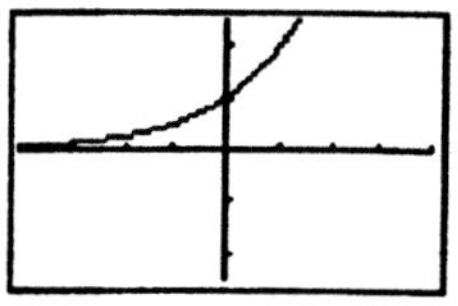

7. $y = 2^x$ _____

8. $y = 3^{-x}$ _____

9. $y = \log_9 x$ _____

10. $y = \log_{1/5} x$ _____

9.3

11. Write $\log_5\left(\frac{\sqrt[3]{x^2y}}{w^2}\right)$ as a sum or difference of multiples of single logarithms.

12. Write $4\log_5(x+3) - 3\log_5 w$ as a single logarithm expression.

For problems 13-15 find the value of each expression.

13. $4^{\log_4 20}$

14. $\log_{(1/5)}(5)^x$

15. $\log_2\left(\frac{1}{16}\right)$

9.4

For problems 16-21 find the value of each expression rounded to four significant digits.

16. log 428

17. log 0.0652

18. log 39400

19. $\frac{\log 72.3}{\log 56.2}$

20. $\frac{\log 672}{\log 0.36}$

21. $\frac{\log 783}{\log 4925}$

For problems 22-24 find N, the antilogarithm, rounded to three significant digits.

22. $\log N = 0.8724$

23. $\log N = 3.7409$

24. $\log N = -2.765$

9.5

For problems 25-26 evaluate each to four significant digits.

25. $e^{3/7}$

26. ln 829

For problems 27-28 find x rounded to three significant digits.

27. $\lambda n\ x = 2.739$

28. $\lambda n\ x = -1.384$

29. It has been determined that a particular bacteria grows at a rate of 12.0% per day. How large will a culture be in 5 days if its initial size is $14\bar{0}0$?

30. A medicine loses its potency at the rate of 8% per month. If its initial potency is 1000 units,

what will be its potency after 4 months? Round to three significant digits.

31. Find the value of $\log_5 3748$ rounded to four significant digits.

9.6

For problems 32-34 solve for x. Round your answers to three significant digits.

32. $3^{4x+2} = 8$

33. $6^{2x/3} = 47$

34. $7e^{5x} = 125$

9.7

For problems 35-36 solve for x. Round your answers to three significant digits.

35. $\log(x+3) + \log 24 = 2.324$

36. $\ln x - \ln(x-5) = \ln 12$

9.8

37. How long will it take an initial investment of \$5000 to grow to \$20,000 if it is invested at 9.6% compounded continuously?

38. The formula for the intensity level of sound is $\beta = 10\log\frac{I}{10^{-16}}$. Compute the intensity of sound in watts per cm^2 for a band whose sound, β, is $14\bar{0}$ dB.

9.9

39. Plot the graph of $y = 5x^2$ on logarithmic graph paper.

40. Plot the graph of $y = 3^x$ on semi-logarithmic graph paper.

Chapter 10 Test

10.1

1. Find the smallest positive coterminal angle in standard position for 135E.

2. Find the largest negative coterminal angle in standard position for 215E.

3. Find the values of sin θ, cos θ, and tan θ if θ is in standard position and its terminal side passes through the point (–7,6). Give exact answers.

4. Find the value of csc θ, sec θ, and cot θ if θ is in standard position and its terminal side passes through the point (5,–8). Give exact answers.

10.2

For problems 5-7, find the reference angle.

5. 125E 6. 248E 7. 295E

For problems 8 and 9, find the value of each trigonometric function, rounded to four significant digits.

8. sin 204.8E 9. sec(–136.4E)

For problems 10-12 find θ for 0E ≤ θ ≤ 360E rounded to the nearest tenth of a degree.

10. sin θ = 0.9542 11. cot θ = –0.9930 12. sec θ = –1.2813

10.3

13. Change 2.47 radians to degrees rounded to the nearest tenth of a degree.

14. Find the reference angle in radians for the angle measuring $\frac{14\pi}{15}$. Give an exact answer.

For problems 15 and 16 give the exact value.

15. $\cos\frac{7\pi}{6}$ 16. $\cot\frac{5\pi}{3}$

17. Find the value of cos(–8.25) rounded to four significant digits.

18. Find θ for $0 \le \theta \le 2\pi$ rounded to three significant digits, given sin θ = –0.653421.

10.4

19. The angular displacement of a revolving body is given by the formula $\theta = \omega t$. The flywheel of a steam engine is rotating at $54\bar{0}$ rpm. Find its angular displacement in 8.00 s.

20. An airplane propeller whose blades are 5.50 feet long is rotating at $24\bar{0}0$ rpm. Find the linear speed of a point on the end of the blade.

Chapter 11 Test

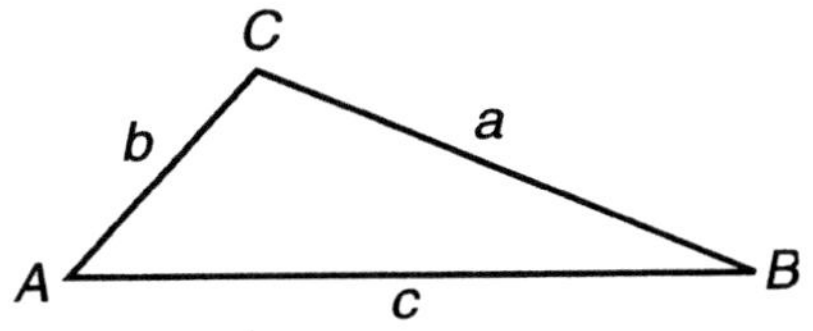

11.1

For problems 1-2 solve the triangles. Express lengths of sides to three significant digits and angles to the nearest tenth of a degree.

1. $A = 32.6$E
   $a = 328$ ft
   $b = 253$ ft

2. $B = 38.7$E
   $b = 97.8$ m
   $c = 72.5$ m

11.2

For problems 3-5 solve the triangles. Express the lengths of sides to three significant digits and the angles to the nearest tenth of a degree.

3. $B = 30.2$E
   $b = 14.3$ cm
   $c = 27.3$ cm

4. $A = 51.6$E
   $a = 94.5$ cm
   $b = 128$ cm

5. $A = 46.5$E
   $a = 36.2$ yd
   $b = 48.5$ yd

11.3

For problems 6-8 solve the triangles. Express the lengths of sides to three significant digits and the angles to the nearest tenth of a degree.

6. $A = 47.3$E
   $b = 92.5$ cm
   $c = 81.7$ cm

7. $a = 245$ ft
   $b = 218$ ft
   $c = 193$ ft

8. $C = 65.2$E
   $a = 427$ m
   $b = 586$ m

11.4

9. Surveyors were interested in finding the distance across an area of quicksand. From a chosen point on dry land they measured the distance to each side of the quicksand, finding the lengths to be 234 ft and 309 ft. The angle between these two lines of measurement was 96.5E. Find the distance across the quicksand.

10. A vertical power line tower stands on the side of a hill which makes an angle of 23.7E with the horizontal. Guy wires are attached $14\bar{0}$ ft up the tower. What lengths of guy wires are needed to reach points 45.0 ft uphill and 92.5 ft downhill from the base of the tower?

11. From a cliff 385 ft above the sea a coastguard officer observed a sinking boat at an

angle of depression of 21.0E. Closer to shore she saw a coastguard cutter at an angle of depression of 38.0E. Find the distance between the coastguard cutter and the sinking boat.

<u>11.5</u>

12. Find the sum of the given vectors by using the parallelogram method. Choose a suitable scale. Angles are in standard position.

$\vec{v}$ = 4.0 m at 18E
$\vec{w}$ = 6.0 m at 71E

<u>11.6</u>

13. Find the sum of the given vectors by using trigonometry. Angles are in standard position.

$\vec{v}$ = 37.0 ft at 105E
$\vec{w}$ = 43.0 ft at 215E

<u>11.7</u>

14. Find the horizontal and vertical components of the vector given in standard position.

$\vec{v}$ = 872 mi at 224E

15. Find the vector whose horizontal and vertical components are:

$v_x = 283$ m $\qquad v_y = -358$ m

16. Use the component method to find the sum of the vectors.

$\vec{v}$ = 72.3 ft at 218E
$\vec{w}$ = 65.8 ft at 294E

17. A series circuit contains a resistance of 253 Ω and a capacitive reactance of 165 Ω. Find the impedance and the phase angle.

<u>11.8</u>

18. An airplane is flying on a heading of 23.0E east of north at an air speed of 285 mph. If the wind is from the south at 34.0 mph, find the plane=s course and ground speed.

19. Sung Ae applies a 34.0-lb force to the handle of a warehouse cart, which makes a 28.0E angle with the ground. Find the horizontal and vertical components of this force.

20. A 975-lb weight hangs from two cables of equal lengths suspended from a support beam as shown. Find the tension in each rope.

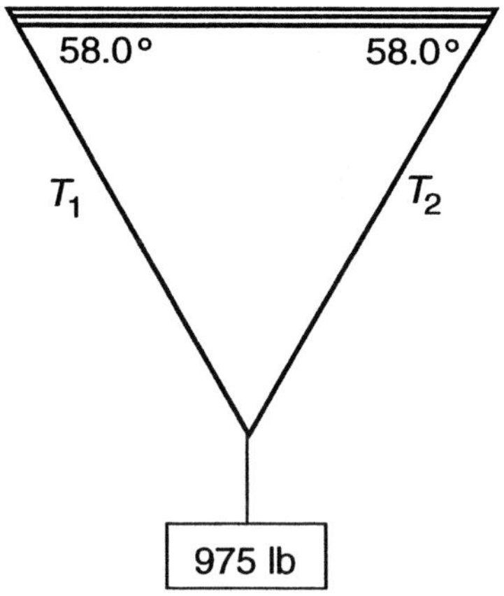

Chapter 12 Test

12.1

Find the amplitude and the period of each function. Sketch each graph through at least two cycles.

1. $y = -4 \cos x$
2. $y = 2 \sin 2x$
3. $y = \frac{1}{2} \sin 4\pi x$
4. $y = 5 \cos 12x$

12.2

Find the amplitude, period, and phase shift for each function, and sketch its graph.

5. $y = \sin\left(x - \frac{\pi}{4}\right)$
6. $y = 2\sin\left(4x + \frac{\pi}{6}\right)$
7. $y = 4\cos\left(6x - \frac{\pi}{2}\right)$

12.3

Find the period of each function and sketch its graph.

8. $y = \tan 6x$
9. $y = 3 \csc 2x$
10. $y = 3 \sec 4x$
11. $y = \cot\left(x - \frac{\pi}{6}\right)$

12.4

Using the method of addition of ordinates, graph each equation for one complete period of the composite curve. Use your graphing calculator to check your work.

12. $y = 2 \sin x + \cos 3x$
13. $y = \frac{1}{2} \cos 2x - \sin 4x$

12.5

14. A crankshaft 2 feet long starts from a position of 45E and rotates in a counterclockwise direction at a uniform angular velocity of 6 revolutions per second (12 π rad/s). Plot the curve that shows the simple harmonic motion of the crank handle, and find the equation of this curve.

15. A rotating wire through a magnetic field generates an alternating current where the maximum instantaneous voltage is 220 V, the angular velocity of the rotating wire is 60 cycles/s, and the phase shift is 0E. Find the equation for this curve.

Chapter 13 Test

13.1

Prove each identity.

1. $\dfrac{\sin^2 x}{1-\cos x} = 1+\cos x$

2. $\dfrac{1+\tan^2 x}{\csc^2 x} = \tan^2 x$

3. $\dfrac{\sec x - \cos x}{\tan x} = \sin x$

4. $\dfrac{\sec^4 x - 1}{\tan^2 x} = \sec^2 x + 1$

13.2

Prove each identity.

5. $\sin\left(x+\dfrac{3\pi}{2}\right) = -\cos x$

6. $\tan\left(x-\dfrac{\pi}{4}\right) = \dfrac{\tan x - 1}{\tan x + 1}$

Simplify each expression.

7. $\sin M \cos N - \cos M \sin N$

8. $\cos(M+N) - \cos(M-N)$

13.3

Simplify each expression.

9. $\cos^2 2\theta - \sin^2 2\theta$

10. $50 \sin 15t \cos 15t$

11. Find the exact value of $\cos 2\theta$ when $\sin\theta = \dfrac{-5}{13}$ and $\pi < \theta < \dfrac{3\pi}{2}$

12. Find the exact value of $\tan\dfrac{\theta}{2}$ when $\tan\theta = \dfrac{-15}{8}$ and $\dfrac{\pi}{2} < \theta < \pi$.

13. Prove the following identity.

$$\tan\frac{\theta}{2} = \frac{\sin\theta}{1+\cos\theta}$$

13.4

Solve each trigonometric equation for $0E \le x < 360E$.

14. $2 \sin 2x + 1 = 0$

15. $2 \sin^2 x + \cos x = 1$

16. Solve for x where $0E \le x < 360E$. Round your answers to the nearest tenth of a degree.

$$4 \cos x + \sin^2 \frac{x}{2} = 2 \cos^2 x + 2$$

13.5

17. Solve for x. $y = 5 \sin(x - 2)$

Find the value of each expression in radians.

18. $\arccos\left(-\frac{\sqrt{3}}{2}\right)$

19. $\sec\left[\arcsin\left(-\frac{1}{2}\right)\right]$

20. Find an algebraic expression for $\sin[\arccos x]$.

Chapter 14 Test

14.1

Perform the indicated operations and simplify.

1. j^{86} 2. $(3 - 7j)(2 + 5j)$ 3. $\dfrac{5+3j}{2-j}$

4. Solve the equation: $3x^2 - 5x + 4 = 0$

14.2

Write each complex number in trigonometric form. Give angles to the nearest degree.

5. $7 - 7j$ 6. $2 + \sqrt{3}j$ 7. $-3 + 8j$

Write each complex number in rectangular form. Round approximate values to three significant digits.

8. $5(\cos 210° + j \sin 210°)$ 9. $3\sqrt{2}(\cos 130° + j \sin 130°)$

10. $8 \underline{/ 295E}$

Write each complex number in exponential form. Round approximate values to three significant digits.

11. $4 - 3j$ 12. $-8 - 8j$ 13. $-6 + 10j$

Write each complex number in rectangular form. Round approximate values to three significant digits.

14. $4e^{1.25j}$ 15. $6e^{3.75j}$ 16. $7e^{5.34j}$

14.3

Find each product and write the result in exponential and rectangular form. Round approximate values to three significant digits.

17. $3e^{2j}(7e^{1.4j})$

18. $5e^{4.2j}(2e^{3.15j})$

19. $e^{5.4j}(4e^{2.65j})$

Find each product and write the result in trigonometric form.

20. $[6(\cos 23° + j\sin 23°)][7(\cos 68° + j\sin 68°)]$

21. $[4(\cos 142° + j\sin 142°)][5(\cos 320° + j\sin 320°)]$

Find each quotient and write the result in exponential and rectangular form. Round approximate values to three significant digits.

22. $\dfrac{48e^{4.5j}}{12e^{2.7j}}$

23. $\dfrac{35e^{1.37j}}{5e^{5.92j}}$

24. $\dfrac{6e^{-2.7j}}{14e^{4.6j}}$

Find each quotient and write the results in trigonometric form. Round approximate values to three significant digits.

25. $\dfrac{64(\cos 307° + j\sin 307°)}{4(\cos 123° + j\sin 123°)}$

26. $\dfrac{28(\cos 104° + j\sin 104°)}{35(\cos 349° + j\sin 349°)}$

14.4

Find each power and write the result in trigonometric and rectangular form. Round approximate values to three significant digits.

27. $(5\,\angle\,40°)^3$

28. $(3\,\angle\,135°)^5$

29. $(2-5j)^4$

30. Find the fourth roots of $1-j$. Write your answer in trigonometric form.

Chapter 15 Test

15.1

1. Give the order of the matrix $\begin{bmatrix} 1 & 2 & 3 & 4 \\ 5 & 6 & 7 & 8 \end{bmatrix}$.

2. Find the values of the variables that make the matrices equal.

$$\begin{bmatrix} x+3 & x+7 \\ 6 & x \end{bmatrix} = \begin{bmatrix} y & 2 \\ y+4 & -5 \end{bmatrix}$$

For problems 3-5 use the given matrices to perform the indicated operations.

$$A = \begin{bmatrix} 3 & -10 \\ -1 & 7 \end{bmatrix}, B = \begin{bmatrix} 8 & -5 & -3 \\ -2 & 6 & 9 \end{bmatrix}, C = \begin{bmatrix} -5 & -1 \\ 4 & 3 \end{bmatrix}$$

3. $A + C$

4. $3A$

5. $2B + C$

15.2

For problems 6-8 find the indicated products.

6. $\begin{bmatrix} 3 & -1 \\ -2 & 5 \end{bmatrix}\begin{bmatrix} 4 & 0 & 10 \\ 8 & 7 & -6 \end{bmatrix}$

7. $\begin{bmatrix} -2 & 6 & 3 \end{bmatrix}\begin{bmatrix} 7 \\ 8 \\ -5 \end{bmatrix}$

8. $\begin{bmatrix} -2 & 3 & 7 \\ 0 & 1 & 4 \\ 5 & -8 & 0 \end{bmatrix}\begin{bmatrix} 6 & 0 & -2 \\ -4 & 7 & 1 \\ 3 & 5 & 8 \end{bmatrix}$

15.3

For problems 9-10 find the inverse matrix.

9. $\begin{bmatrix} 3 & 4 \\ 1 & 2 \end{bmatrix}$

10. $\begin{bmatrix} 5 & -8 & 0 \\ -2 & 3 & 0 \\ 0 & 1 & 4 \end{bmatrix}$

11. Use the matrix method to solve the following system of equations.

$$\begin{aligned} x+y-z&=-9 \\ 2x-y-2z&=-3 \\ x+2y+3z&=14 \end{aligned}$$

12. A company makes three types of backpacks for a cost of \$754 for 100 backpacks. Production costs for the three types are \$5, \$8, and \$12. They sell for \$20, \$30, and \$50 respectively. How many of each type are made if the profit is \$2226?

Chapter 16 Test

16.1

For problems 1-2 find the degree of each polynomial.

1. $3x^8 + 5x^4 - 6x^7 - x^9$

2. $4x^6 - 5x^3 + 2x^7 - 16$

For problems 3-4 find the x and y intercepts of each function.

3. $f(x) = (x - 5)(3x - 2)^2$

4. $f(x) = x^3 - x^2 - 8x + 12$

5. Given that a third degree polynomial has zeros of $\frac{1}{2}$, $\frac{7}{3}$, and -5, write the polynomial function in the form $f(x) = ax^3 + bx^2 + cx + d$ where the coefficients are integers.

6. Name the zeros of $f(x) = 5(x - 7)^3(x + 4)^2$ and the multiplicity of each.

7. Given the graph of $g(x)$, is the zero at -3 of odd or even multiplicity?

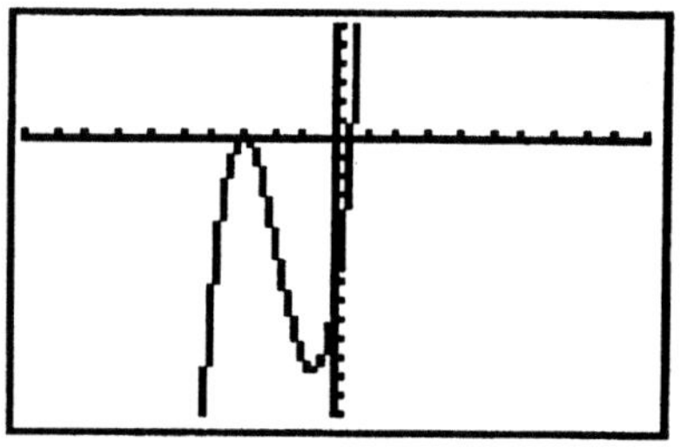

16.2

For problems 8-9 solve for x. Round your answer to four significant digits.

8. $x^3 + 5 = 7x$

9. $x^3 = 8x^2 + 2x - 4$

16.3

10. Given that $3 + j$ is a solution of the equation $x^4 - 9x^3 - 12x^2 + 210x - 400 = 0$, find all of its complex solutions.

11. Find a polynomial equation with real coefficients and degree 5, whose solutions include -7 and $4 - 5j$.

12. Find all complex roots of the equation $x^4 - x^3 + 7x^2 - 9x - 18 = 0$. *Hint*: Start by using the graph.

Chapter 17 Test

17.1

For problems 1-3 use an inequality to express those real numbers shaded on each number line.

1. 

2. 

3. 

For problems 4-6 solve each inequality.

4. $-2.5x < 20$

5. $3(2x - 5) \geq 9x + 15$

6. $\frac{x}{5} > -15$

17.2

7. Solve the equation $*2x - 7* = 13$.

For problems 8-10 solve each inequality. Graph the solution on the number line.

8. $*x + 3* \leq 8$

9. $*x - 7* > 5$

10. $*5 - 2x* \leq 13$

17.3

For problems 11-13 solve each inequality.

11. $(2x - 5)(x + 7) < 0$

12. $\frac{x+6}{2x-10} \geq 0$

13. $\frac{4-x}{x} \leq 3$

17.4

For problems 14-16 graph each inequality.

14. $y \geq 2x - 7$

15. $y < 4$

16. $4x - 8y \geq 15$

For problems 17-18 solve each system of inequalities graphically.

17. $y \geq x^2 - 4$
$y \leq 2$

18. $y \leq 2x + 5$
$y \geq 0$
$x \geq 0$

Chapter 18 Test

18.1

For problems 1-4, find the *n*th term of the arithmetic progression.

1. $5, 8, 11, \ldots, n = 7$
2. $2, -1, -4, \ldots, n = 8$
3. $-1, 0.5, 2, \ldots, n = 17$
4. $7, 4.5, 2, \ldots, n = 19$

For problems 5-7, find the sum of the first *n* terms.

5. $a = 3,\ \lambda = 23,\ n = 6$
6. $a = 3,\ \lambda = -9,\ n = 7$
7. $12, 7, 2, \ldots, n = 17$

8. Carmen is employed at an initial salary of \$32,000. If she receives an annual raise of \$1100, what is her salary for the eighth year?

18.2

For problems 9-12, find the *n*th term of the geometric progression.

9. $4, 12, 36, \ldots, n = 5$
10. $2, \frac{1}{2}, \frac{1}{8}, \ldots, n = 7$
11. $-9, -3, -1, \ldots, n = 8$
12. $-5, -10, -20, \ldots, n = 10$

For problems 13-15, find the sum of the first *n* terms.

13. $a = 3,\ r = \frac{1}{2},\ n = 4$
14. $a = 2, r = 3, n = 6$
15. $81, 27, 9, \ldots, n = 6$

16. A sugar syrup is being cooled at the rate of 25% each minute. Find the temperature of the syrup after 6 minutes if the original temperature was 512EF. (*Hint*: The temperature after 1 minute is 384EF.)

For problems 17-19, find the sum of the infinite geometric series, if possible.

17. $15+3+\frac{3}{5}+\ldots+15\left(\frac{1}{5}\right)^{n-1}+\ldots$

18. $12+2+\frac{1}{3}+\ldots+12\left(\frac{1}{6}\right)^{n-1}+\ldots$

19. $\frac{1}{2}+1+2+\ldots+\frac{1}{2}(2)^{n-1}+\ldots$

18.3

For problems 20-22, expand each binomial using the binomial theorem.

20. $(x+4y)^4$

21. $(2x-5)^6$

22. $(4x-3y)^5$

For problems 23-24, find the indicated term of each binomial expansion.

23. $(5x-2y)^7$ fourth term

24. $(3x-5y)^6$ third term

Chapter 19 Test

19.1

1. Hybrid seed corn #5 was tested on 40 farms. The average harvest in bushels per acre is given below. Construct a frequency table, a histogram, and a frequency polygon using 9 intervals each containing 4 numbers.

170	148	171	161	156
183	172	168	149	173
162	153	162	158	155
152	167	150	175	182
162	171	183	173	148
181	173	169	162	181
165	179	173	171	149
176	166	156	155	160

19.2

2. Find the mean, median, and mode for the data set {29, 24, 32, 28, 25, 24, 31, 30}. Round to three significant digits.

3. Find the mean, median, and mode for the following data set. Round to three significant digits.

Measurement	24	25	26	27	28
Frequency	8	14	15	19	10

19.3

For problems 4-5, find the range and sample standard deviation. Round to three significant digits.

4. {3, 5, 6, 6, 8, 9, 11}

5. {29, 32, 33, 36, 38}

19.4

6. The mean and sample standard deviation are given. Find the percent of data points a) within one sample standard deviation of the mean and b) within two sample standard deviations of the mean. Are these percents within two of the $\bar{x} \pm 1s_x$ and $\bar{x} \pm 2s_x$ criteria for normal distribution?

Metres	17	19	21	23	25	27	29
Frequency	4	17	32	39	29	18	4

$\bar{x} = 23.0$ $s_x = 2.77$

19.5

7. Use the matrix method to find the line of best fit, $y = ax + b$. Round to two significant digits. Check your answer by using a calculator.

x	0	1	2	3	4	5
y	3	1	−2	−4	−7	−9

8. Use the matrix method to find the quadratic equation of best fit, $y = ax^2 + bx + c$. Round to two significant digits. Check your answer by using a calculator.

x	−1	0	1	2	3	4
y	9	7	3	−3	−12	−20

9. Use a calculator program to find the exponential curve of best fit, $y = ab^x$, for the chart of bacteria growth.

x	1	2	3	4	5	6
y	15	109	805	5900	45000	325000

10. A process is in control with control limits 17.996 mm to 18.004 mm. Determine its capability to produce an order with specifications 18 ± 0.005 mm.

Chapter 20 Test

20.1

1. Graph the circle with center at $(-2, -3)$ and $r = 4$.

2. Find the equation of the circle (in standard form) with center at $(4, -1)$ and radius 3.

3. Find the equation of the circle (in standard form) with center at $(7, 3)$ and passing through $(6, -2)$.

4. Find the center and radius of the given circle. $x^2 + y^2 - 6x + 4y - 3 = 0$

20.2

5. Find the focus and the directrix of the parabola $y^2 = -20x$.

6. Find the equation of the parabola with focus $F(0, -3)$ and directrix $y = 3$.

7. Find the equation of the parabola with vertex at $(0, 0)$ and directrix $y = -6$.

8. Find the vertex and equation of the axis of the parabola $y = 2x^2 - 7x - 15$.

20.3

For problems 9-10, find the vertices, foci, and lengths of the major and minor axes of each ellipse.

9. $\frac{x^2}{9} + \frac{y^2}{64} = 1$

10. $25x^2 + 144y^2 = 3600$

11. Find the equation of the ellipse with vertices at $(\pm 7, 0)$ and foci at $(\pm 2\sqrt{10}, 0)$.

12. Find the equation of the ellipse with vertices at $(0, \pm 8)$, and with a minor axis of length 10.

20.4

For problems 13-14, find the vertices, foci, and lengths of the transverse and conjugate axes for each hyperbola. Find the equations of the asymptotes.

13. $\dfrac{x^2}{36} - \dfrac{y^2}{64} = 1$

14. $\dfrac{y^2}{100} - \dfrac{x^2}{81} = 1$

15. Find the equation of the hyperbola with vertices $(\pm 9, 0)$ and foci at $(\pm 12, 0)$.

16. Find the equation of the hyperbola with foci at $(0, \pm 7)$ and transverse axis of length 10.

20.5

For problems 17-18, find the equation of each curve from the given information.

17. Parabola with vertex at (3, 4), focus at (3, 6) and directrix $y = 2$.

18. Ellipse with center at (–3, 7), vertices at (–3, 20) and (–3, –6) and minor axis of length 10.

19. Name the graph of the equation $4x^2 - 9y^2 + 8x + 36y - 68 = 0$. Locate the vertex if it is a parabola or the center if it is an ellipse or hyperbola.

20.6

20. Determine whether the following equation represents a circle, a parabola, an ellipse, or a hyperbola. $x^2 + y^2 + 10x - 6y - 15 = 0$

20.7

21. Use the substitution method to find the real solutions for the following system of equations. Round to two significant digits.

$$\frac{x^2}{16} - \frac{y^2}{25} = 1 \text{ and } (x-3)^2 + y^2 = 9$$

20.8

22. For the point (4, 45E), name three other sets of polar coordinates such that $-360° \le \theta \le 360°$.

23. Graph by plotting points. $r = 6 - 6 \sin \theta$

24. Change (5, 60E) to rectangular coordinates.

25. Change $(-4\sqrt{2}, -4\sqrt{2})$ to polar coordinates in radians.

26. Change $x^2 + y^2 = 49$ to polar form.

27. Change $r = \dfrac{-5}{1+\sin\theta}$ to rectangular form.

20.9

For problems 28-32, graph each equation.

28. $r = 7 \sin \theta$

29. $r = 8 + 4 \cos \theta$

30. $r = 5 + 12 \sin \theta$

31. $r = 3 \sin 5\theta$

32. $r = 6 \cos 4\theta$

Chapter 21 Test

21.1

For problems 1-2, evaluate each function at the given value.

1. $f(t) = 2t^3 + t - 6;\ t = 4$

2. $f(x) = \sqrt{x^2 - 4x};\ x = 5$

For problems 3-4, find a) Δs and b) $\bar{v}$ for each function expressing distance s in terms of time t.

3. $s = 5t + 3$

4. $s = 3t^3 - 10t + 3$

5. Given $s = 4t^2 + 16t - 5$, find $\bar{v}$ at $t = 3$ and $\Delta t = 1$, where s is in metres and t is in seconds.

6. Given $s = f(t) = 4t^2 - 9$, find the instantaneous velocity v of an object moving along a straight line when $t = 3$. s is in meters and t is in seconds.

21.2

For problems 7-12, find each limit.

7. $\lim\limits_{x \to 5} \dfrac{x^2 - 25}{x - 5}$

8. $\lim\limits_{x \to \infty} \dfrac{4x^2 - 7}{8x^2 + 5}$

9. $\lim\limits_{x \to 8} 5$

10. $\lim\limits_{x \to -3/2} \dfrac{2x^2 - 9x - 18}{2x + 3}$

11. $\lim\limits_{x \to \infty} \dfrac{1}{x^3}$

12. $\lim\limits_{x \to 2} (3x^2 - 2x)$

21.3

13. Find the slope of the line tangent to the curve $y = x^2 - 4x + 7$ at (5, 12).

14. Find the equation of the line tangent to the curve $y = 3x^2 - 9$ at (4, 39).

15. Find the point or points on the curve $y = 2x^3 - 5x^2$ where the slope of the tangent line is 0.

21.4

For problems 16-17, find the derivative of each function.

16. $y = 2x^2 - 5x$

17. $y = \dfrac{1}{2x+7}$

18. Find the equation of the line tangent to $y = x^3$ at the point (1, 8).

21.5

For problems 19-23, find the derivative of each function.

19. $y = x^{14}$

20. $y = 2x^4 - 3x^2 + 7$

21. $y = \sqrt{6}x^4$

22. $y = \dfrac{2}{x^8}$

23. $y = \dfrac{1}{x^2\sqrt{x}}$

24. For $f(x) = 3x^4 - x^2 + 5$, find $f'(2)$.

21.6

For problems 25-28, find the derivative of each function.

25. $y = (x^4 - 2x^3)(x^3 + 5x)$

26. $x^8(3x^2 + 5x - 7)$

27. $y = \dfrac{x}{4x-9}$

28. $y = \dfrac{2x+3}{3x^2-5}$

21.7

For problems 29-31, find the derivative of each function.

29. $y = (3x - 7)^9$

30. $y = \dfrac{2}{(x^3-8)^5}$

31. $y = \dfrac{(4x-1)^6}{(3x+2)^4}$

21.8

For problems 32-35, find the derivative using implicit differentiation.

32. $8x + 3y = 6$

33. $9x^2 + 4y^2 = 36$

34. $\dfrac{2x + y}{x - 3y} = x^3$

35. $y = 3xy^2 + \sqrt{x - 1}$

36. Find the equation of the line tangent to the curve $3xy - 5y^2 = 10$ at (5, 2).

21.10

37. Given $y = 4x^5 - 3x^4 + 8x^2$, find the first three derivatives of y.

38. Find y' and y'' implicitly and express the result in x and y. $(2y + 3)^2 = 8x$

Chapter 22 Test

22.1

In problems 1-4, determine the intercepts, the intervals in which the curve is above or touching the x-axis, symmetry, and asymptotes. Graph each equation.

1. $y^2 = \frac{1}{x-3}$

2. $y = x^3 - 5x^2 - 12x + 36$

3. $y = \frac{4}{5x-3}$

4. $y = \frac{x}{(x+4)(x-5)}$

22.2

In problems 5-6, find any relative maximum or minimum values of each function and sketch the curve.

5. $y = x^3 + 2x^2 - 7x + 4$

6. $y = x^4 - 6x^2 + 5$

22.3

7. Given $f(x) = \frac{6}{x^2 - 9}$, find a) the intervals for which $f(x)$ is increasing and decreasing, b) relative maximums and relative minimums, c) intervals for which $f(x)$ is concave upward and concave downward, and d) points of inflection. Sketch the curve.

22.4

In problems 8 and 9 use a graph to find estimates for the real solutions of each equation, then use Newton's Method to improve your estimates to at least six significant digits. For the trip equation, be sure your calculator is set in radians.

8. $x^4 - 4x^2 - 21 = 0$

9. $3 \sin x = \cos x, \quad 0 \leq x < 2\pi$

22.5

8. A rectangular box, open at the top, with a square base is to have a volume of 13,500 cm^3. Find the dimensions if the box is to contain the least amount of material.

9. The power P in watts (W) in a circuit with resistance R in ohms (Ω) varies according to $P = \dfrac{49R}{(R+9)^2}$. Find the resistance that gives the maximum power.

22.6

10. Given $y = 5x^3$ and $\dfrac{dx}{dt} = 3$ at $x = 2$, find $\dfrac{dy}{dt}$.

11. A circular plate is being heated so that its radius is increasing at a rate of 0.09 in./h. How fast is its area increasing when its radius is 15 inches?

12. A cylinder with an inside radius of 35 mm is sealed at one end and a piston is at the other end. At what rate is the piston moving if fluid is being pumped into the cylinder at the rate of 60 cm^3/s? *Hint*: Pay attention to units.

22.7

13. Find dy for the expression $9x^2 + 25y^2 = 10$.

14. Using a differential expression, find the change in $V = \dfrac{4}{3}\pi r^3$ from $r = 25.00$ in. to 25.10 in.

15. Find the percentage error in problem 14.

16. The town of Darlington plans to build a spherical water tower with an inner diameter of 30.00 m and sides of thickness 5.00 cm. Find the approximate volume of steel needed using differentials. If the density of steel is 7800 kg/m^3, find the approximate amount of steel needed.

Chapter 23 Test

23.1

For problems 1-4, find the derivative of each function.

1. $y = \cos 15x$

2. $y = 3\sin(2x + 1)$

3. $y = \dfrac{3x^2 + 5}{\sin 2x}$

4. $y = \sin 6x \cos 8x$

23.2

For problems 5-8, find the derivative of each function.

5. $y = \tan\sqrt{x^2 + 4x}$

6. $y = \dfrac{\sec x}{x^2}$

7. $y = \sin 3x \cot^2 x$

8. $y = (1 + \csc x)^3$

23.3

For problems 9-12, find the derivative of each.

9. $y = 5\text{ Arc}\sin(x - 1)$

10. $y = 6\text{Arc}\cot\left(\dfrac{x}{2}\right)$

11. $y = 4\text{Arc}\cos(1 - x^2)$

12. $y = \text{Arc}\csc^4 2x$

23.4

For problems 13-14, find the derivative of each function.

13. $y = \log(8x + 2)$

14. $y = \ln(5x^2 - 15)$

For problems 15-16, find $\frac{dy}{dx}$ using logarithmic differentiation.

15. $y = \dfrac{(x-3)^2(5x+4)}{(2x-7)(6-x)}$

16. $y = (\cos x)$

23.5

For problems 17-19, find the derivative of each function.

17. $y = e^{x^3+3x}$

18. $y = \sin e^{2x}$

19. $y = x^2 e^x$

23.6

Find the indicated limits, if they exist.

20. $\lim\limits_{x \to 1} \dfrac{\ln x}{x^2 - 1}$

21. $\lim\limits_{x \to 0} \dfrac{x^4}{1 - \cos x}$

23.7

20. Find any relative maximums or minimums of the function $y = x \ln(x^2)$.

Chapter 24 Test

24.1

For problems 1-4, integrate.

1. $\int 3x^{23}dx$

2. $\int (3x+1)^{1/2}dx$

3. $\int \frac{x^2dx}{(4x^3-5)^3}$

4. $\int \frac{(x^4+2x^2)dx}{\sqrt[4]{3x^5+10x^3}}$

24.2

5. Find the equation of the curve $y=f(x)$ satisfying the conditions: $\frac{dy}{dx}=(x^2-6)^3x\,dx$ and passing through (2, 5).

6. A ball is hurled straight up from the top of a 28 ft building at a velocity of 44 ft/s. a) Find the equation describing the altitude of the ball from the ground. b) How long does it take for the ball to hit the ground? Round to three significant digits.

7. A capacitor with a capacitance 10^{-3} F has a voltage of 120 V across it. At a given instant ($t=0$) the capacitor is connected to a source that sends a current $i=6\sqrt{t}+.01A$ through the circuit. Find the voltage across the capacitor when $t=0.36$ s.

24.3

For problems 8-9, find the area under the curve.

8. $y=-x^2+3x+10$ from $x=1$ to $x=4$

9. $y=\sqrt{4x+1}$ from $x=2$ to $x=6$

10. Find the area bounded by the x-axis and $y=-x^2+10x-16$.

24.4

For problems 11-14, evaluate each definite integral.

11. $\int_{2}^{3}(6x^2 - x - 6)dx$

12. $\int_{-3}^{0}2x\sqrt{25 - x^2}\,dx$

13. $\int_{0}^{4}\frac{18x\,dx}{\sqrt{9x^2 + 25}}$

14. $\int_{2}^{5}(x^3 - 9x^2 + 6x - 3)dx$

Chapter 25 Test

25.1

For problems 1-2 find each area bounded by the curves.

1. $y = 2x^2 - 5x - 8$
   $y = x + 12$

2. $y = x^3$
   $y = 2x$

25.2

3. Use the disk method to find the volume of the solid formed by revolving the region bounded by $y = 0.5x^2$, $x = 0$, and $y = 6$ about the y-axis.

25.3

4. Use the shell method to find the volume of the solid formed by revolving the region bounded by $y = \sqrt{x}$, $y = 0$, and $x = 9$ about the x-axis.

5. Use any method to find the volume of the solid formed by revolving the region bounded by $y = x^2$ and $y = 3x$ in the first quadrant about the y-axis.

25.4

6. There is a mass of 14 at $(-10, 0)$ and a mass of 26 at $(6, 0)$. Find where a mass of 12 should be placed on the x-axis so that $(5, 0)$ is the center of mass.

7. Find the center of mass for the two-dimensional system: $m_1 = 8$ at $(3, 5)$, $m_2 = 5$ at $(-1, -3)$, $m_3 = 20$ at $(-2, 4)$.

25.5

8. Find the center of mass measured from the lighter end of a straight wire 16 cm long and whose density is given by $\rho(x) = 8 + x^2$, where x is the distance from one end.

9. Find the centroid of the region bounded by $y = 9 - x^2$ and $y = 0$.

25.6

10. Find the moment of inertia and the radius of gyration of the region bounded by $y = 12 - x^2$, $y = 3$, and $x = 0$ about the y-axis. $\rho = 4$

11. Find the moment of inertia and the radius of gyration of the solid formed by revolving the region bounded by $y = x^2$, $y = 0$, and $x = 4$ about the x-axis. $\rho = 5$

25.7

12. A vertical gate in a dam is in the shape of an isosceles trapezoid 16.0 ft across the top and 10.0 ft across the bottom, with a height of 12.0 ft. Find the force against the gate if the water surface is at the top of the gate.

Chapter 26 Test

26.1

For problems 1-3, integrate.

1. $\int (x^3 + 2x^2)^8 (3x^2 + 4x)dx$

2. $\int e^{6x} dx$

3. $\int \sin 5x \cos 5x \, dx$

26.2

For problems 4-5, integrate.

4. $\int \frac{(x+1)dx}{3x^2 + 6x}$

5. $\int x^2 e^{x^3} dx$

6. Evaluate $\int_2^4 \frac{6 \, dx}{3x - 5}$

26.3

For problems 7-8, integrate.

7. $\int \sec^2 (5x + 10)dx$

8. $\int \sec 3x \tan 3x \, dx$

26.4

For problems 9-10, integrate.

9. $\int \sin^3 5x \cos^2 5x \, dx$

10. $\int \tan^3 8x \, dx$

26.5

For problems 11-12 integrate.

11. $\int \frac{dx}{\sqrt{5 - 45x^2}}$

12. $\int \frac{dx}{4 + 25x^2}$

26.6

For problems 13-14 write each expression as a sum of partial fractions and integrate.

13. $\int \frac{(8x-29)dx}{(x+2)(x-7)}$

14. $\int \frac{(11x^2-18x+3)dx}{x(x-1)^2}$

26.7

For problems 15-16, integrate.

15. $\int xe^{10x}dx$

16. $\int x\ln x\, dx$

26.8

For problems 17-18, integrate.

17. $\int \frac{dx}{(49-x^2)^{3/2}}$

18. $\int \frac{x^3 dx}{\sqrt{x^2+25}}$

26.9

For problems 19-20 integrate using a table or a graphing utility.

19. $\int \frac{dx}{4x(3+4x)}$

20. $\int e^{2x}\sin 5x\, dx$

26.10

21. Use Simpson's rule with $n = 6$ to find the approximate value of $\int_1^4 \frac{dx}{x+4}$.

26.11

22. Find the area within the inner loop of $r = 4 + 8\sin\theta$.

26.12

For problems 23-24 evaluate the improper integrals.

23. $\int_{-\infty}^{0} e^{4x}\,dx$

24. $\int_{0}^{8} \frac{1}{\sqrt[3]{x}}\,dx$

Chapter 27 Test

27.1

1. Name and sketch the graph of $9x^2 + y^2 + 4z^2 = 36$.

2. Find the distance between the points (4, –5, 8) and (3, 7 –2).

27.2

3. Find $\frac{\partial z}{\partial x}$ and $\frac{\partial z}{\partial y}$ for $z = e^{4x} \cos xy$.

4. Given the surface $z = 36x^2 + 25y^2$ find the slope of the tangent line parallel to the xz-plane and through the point (–1, 1, 61).

27.3

5. Find the total differential for $z = 4x^2 + 2xy + 3y^3$.

6. The total resistance R of two resistors R_1 and R_2, connected in parallel, is $\frac{R_1R_2}{R_1 + R_2}$. R_1 measures 525 Ω with a maximum error of 35 Ω. R_2 measures 375 Ω with a maximum error of 24 Ω. Use a differential to approximate the change in R to two significant digits.

7. Find all critical points and any relative maximum or minimum points or saddle points for the function: $z = 2x^2 + 5y^2 - 8xy + 6x$

27.4

8. Evaluate $\int_1^3 \int_0^{x^2} (x^2 - 5xy)\,dy\,dx$.

9. Find the volume of the solid bounded by the plane $5x + 2y + z = 10$ and the coordinate planes.

Chapter 28 Test

28.1

For problems 1-2 determine whether each series converges or diverges.

1. $1+\frac{1}{8}+\frac{1}{27}+\Lambda+\frac{1}{n^3}+\Lambda$

2. $\sum_{n=1}^{\infty}\frac{1}{(3n-1)^2}$

28.2

For problems 3-4 use either the ratio test or the integral test to determine whether each series converges or diverges.

3. $\sum_{n=1}^{\infty}\frac{n+4}{n\cdot 5^n}$

4. $\sum_{n=1}^{\infty}\frac{n^2}{n^3+1}$

28.3

For problems 5-7 determine whether each alternating series converges or diverges. If it converges, find whether it converges absolutely or converges conditionally.

5. $\sum_{n=2}^{\infty}(-1)^n\frac{\sin n}{n^2}$

6. $\sum_{n=1}^{\infty}(-1)^{n+1}\frac{n^3}{n^3+1}$

7. $\sum_{n=1}^{\infty}(-1)^{n+1}\frac{1}{\sqrt[3]{n}}$

28.4

For problems 8-9 find the interval of convergences of each series.

8. $\sum_{n=1}^{\infty}\frac{(6x)^n}{(3n)!}$

9. $\sum_{n=1}^{\infty}\frac{(-1)^n(x-5)^n}{\sqrt[3]{n}}$

28.5

For problems 10-11, find a Maclaurin series expansion for each function. Include at least three non-zero terms.

10. $f(x)=e^{2x}$

11. $f(x)=\sin 3x$

28.6

12. Find a Maclaurin series expansion for $f(x) = \frac{e^x - 1}{x}$.

13. Evaluate $\int_0^1 \cos\sqrt{x}dx$. (Use three non-zero terms.) Round to four significant digits.

28.7

For problems 14-15 find the Taylor series expansion for each function for the given value of a. Give at least three terms.

14. $f(x) = \frac{2}{x^3}, a = 2$

15. $f(x) = \cos x, a = \frac{\pi}{4}$

28.8

16. Calculate the value of $e^{1.2}$ using the first four non-zero terms of a Taylor series. Round to five significant digits.

28.9

17. Find the Fourier series expansion of $f(x) = 3x, 0 \le x < 2\pi$. Write at least four terms.

Chapter 29 Test

29.1

1. State the order and degree of the differential equation $(y''')^2 - 2xy' + 5y'' = 0$.

2. Verify that the function $y = f(x)$ is a solution of the differential equation

$$x\frac{dy}{dx} - y = x^5. \qquad f(x) = \frac{1}{4}(x^5 + 15x).$$

29.2

3. Solve the differential equation: $x(y - 1)\,dx + (x^2 + 3)\,dy = 0$

4. Find the particular solution to the differential equation $\frac{dy}{dx} = \frac{y^2}{x^3}$ where $y = 8$ when $x = 4$.

29.3

For problems 5-6 solve the differential equation.

5. $\frac{y\,dx}{x^2 + y^2} = \frac{x\,dy}{x^2 + y^2} + y^2 dy$

6. $\frac{x\,dy + y\,dx}{4x^3} = dx$

7. Find the particular solution of the differential equation $\frac{x\,dy - y\,dx}{x^2} = 2y\,dy$ where $y = 6$ when $x = 2$.

29.4

For problems 8-9 solve each differential equation.

8. $\frac{dy}{dx} + 6xy = e^{-3x^2}\sin x$

9. $4y\,dx + x\,dy = 5x^2\,dx$

29.5

10. Radioactive material decays at a rate proportional to the amount present. For a certain radioactive substance, approximately 16% of the original quantity decomposes in 24 years. Find the half-life of this radioactive material.

11. Find the equation for the current i in a series circuit with inductance $L = 0.2$ H, resistance $R = 120\ \Omega$, and voltage $V = 210$ volts if the initial current $i_0 = 5$ amperes when $t = 0$. *Hint*: Use $L\frac{di}{dt} + Ri = V$.

12. An object at 450E is cooled in air, which is at 75EF. If the object is 420EF after 6 min, find the temperature of the object after 45 min.

Chapter 30 Test

30.1

For problems 1-2 solve each differential equation.

1. $y'' + 4y' - 21y = 0$

2. $y'' + 8y' = 0$

3. Find the particular solution subject to the given conditions. $y'' - 4y' - 12y = 0$; $y = 1$ and $y' = -18$ when $x = 0$.

30.2

For problems 4-5 solve each differential equation.

4. $y'' - 22y' + 121y = 0$

5. $y'' - 4y' + 29y = 0$

6. Find the particular solution subject to the given conditions. $y'' + 81y = 0$; $y = 7$ and $y' = -27$ when $x = \dfrac{\pi}{18}$.

30.3

For problems 7-8 solve each differential equation.

7. $y'' + 2y' - 15y = x + e^{2x}$

8. $y'' + y = 4 \sin x$

30.4

9. A spring is stretched 6 in. by a weight of 10 lb. If the weight is displaced a distance of 4 in. from a rest position and then released, find the equation of motion.

10. A spring is stretched 8 in. by a weight of 25 lb. A damping force exerts a force of 6 lb for a velocity of 5 in./s. If the weight is displaced from the rest position and then released, find the general equation of motion.

11. An electric circuit has an inductance $L = 0.4$ H, a resistance $R = 250\ \Omega$, and a capacitance $C = 5 \times 10^{-4}\ F$. Find the equation for the current i. *Hint*:

$$\frac{d^2i}{dt^2} + \frac{R}{L}\frac{di}{dt} + \frac{1}{CL}i = 0$$

<u>30.5</u>

Use a table of Laplace Transforms to find the Laplace transform of each function $f(t)$ in problems 12-13.

12. $f(t) = e^{4t} \cos 8t + 5te^{10t}$

13. $f(t) = \dfrac{t^4}{24} + 1 - \cos 6t$

14. Find the inverse transform for the function $F(s)$. $F(s) = \dfrac{9}{s^2+81} + \dfrac{1}{(s+5)^3}$

<u>30.6</u>

For problems 15-16 solve each differential equation subject to the given conditions by using Laplace transforms.

15. $y'' - 5y' - 24y = 0;\ \ y(0) = 17$ and $y'(0) = 26$

16. $y'' - 8y' + 16y = te^{4t};\ \ y(0) = 0$ and $y'(0) = 0$

Answers to Chapter Tests

CHAPTER 1

1. −8 **2.** −72 **3.** −13 **4.** $3\frac{9}{11}$ **5.** 0 **6.** −4 **7.** 3 **8.** $\frac{3}{8}$ **9.** meaningless **10.** $\frac{1}{10^3}$

11. 3.87×10^6 **12.** 0.0000265 **13.** 6.48×10^{-10} **14.** a) 2384.0 km b) 0.00007 km

15. a) 0.000003 *A* b) 0.057 *A* **16.** 34.95 cm **17.** $5.3 \times 10^{-4} m^3$ **18.** $5x^3 - 9x + 16$

19. $-6x^4 + 4x^3 + 3x^2 - 11x$ **20.** $-\frac{1}{4}$ **21.** 7 **22.** x^{14} **23.** w^6 **24.** t^{12} **25.** $32x^{10}y^{15}$ **26.** 7.14

27. 10.8 **28.** 385,000 **29.** $9x^2 - 16$ **30.** $25x^4 - 30x^2 + 9$ **31.** $\frac{5y^3}{7x^6}$ **32.** $15x^3 - 9x^2 - 6x + 11$

33. $2x^2 - 5x + 7$ **34.** 8 **35.** −8.2 **36.** $Q = XP$ **37.** $x = \frac{aF + w}{h}$ **38.** 130 cm **39.** 63.2 Ω

40. 8.57 lb of 60%, 6.43 lb of 25% **41.** 12.5 kg of 18%, 2.5 kg of 10% **42.** 2.01 **43.** 3.27

44. $31{,}20\bar{0}$ and $39{,}00\bar{0}$ **45.** 1218 strokes **46.** $k = 3.8$, $y = 836$ **47.** $k = 6.5$, $m = 144$

CHAPTER 2

1. parallel lines **2.** complementary angles **3.** vertical angles **4.** 75° **5.** 105°

6. isosceles triangle **7.** altitude (or height) **8.** perimeter **9.** 8.03 cm

10. 27.5 ft^2 **11.** 57.4° **12.** 18.1 ft **13.** a) \$11,550 b) 38 bags at \$11.95 per bag is \$454.10

14. 96.8 m **15.** C = 229 m A = 4190 m^2 **16.** 4.78 cm **17.** 65° **18.** 65° **19.** 32.5°

20. a) 41.1 m^3 b) s = 6.36 m A = 51.9^2 **21.** a) 463,000 ft^3 b) $29{,}\bar{0}00$ ft^2 **22.** a) 4010 cm^3

b) 1440 cm^2

CHAPTER 3

1. 39.6 cm **2.** 8.84 mm **3.** 0.654 **4.** 0.923 **5.** 1.49 **6.** 1.70 **7.** 0.8415 **8.** 0.7931 **9.** 2.647

10. 0.3485 **11.** 0.9474 **12.** 1.159 **13.** 78.24° **14.** 42.58° **15.** 27.92° **16.** 52.74°

17. $b = 34.5$ ft, $A = 49.0°$, $B = 41.0°$ **18.** $B = 68.3°$, $a = 2.56$ cm, $c = 6.92$ cm **19.** 56.7 ft

20. 229 km **21.** 8580 ft **22.** $Z = 158\ \Omega$, $\phi = 37.1°$

CHAPTER 4

1. no, $D = \{-3\}$, $R = \{4, 6, 8, 10\}$ **2.** yes, D = Real Numbers, $R = \{y \mid y \geq -1\}$

3. yes, $D = \{x \mid x \geq 2\}$, $R = \{y \mid y \leq 5\}$ **4.** no, $D = \{x \mid x \leq 7\}$, $R =$ Real Numbers

5. a) 0.5 b) 3.25 c) $\dfrac{1-3a^2}{a}$

6.

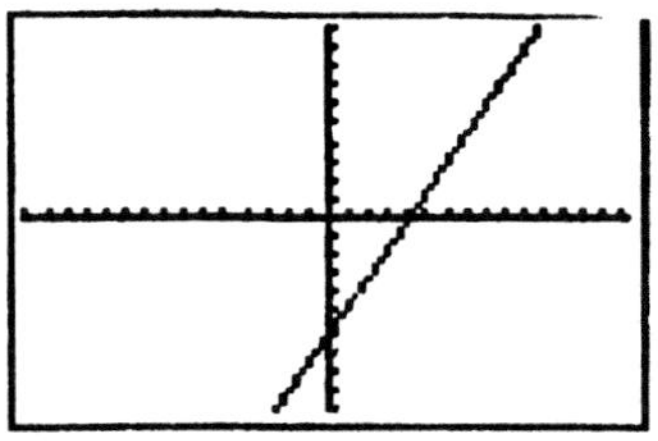

$3x - 2y = 12$

7.

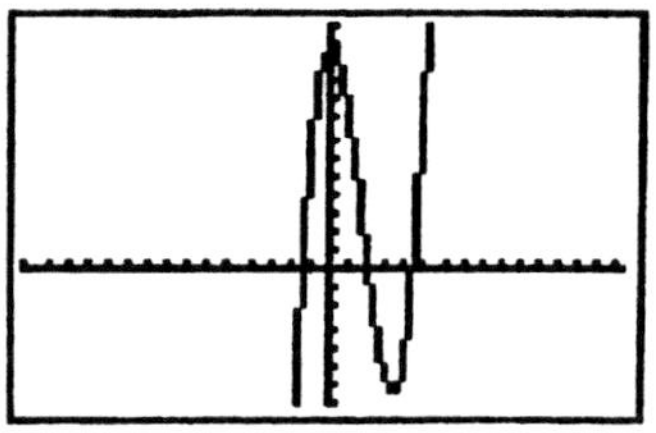

$y = x^3 -$

$5x^2 + 12$

8. $m = \dfrac{-5}{6}$ **9.** $m = \dfrac{-5}{3}$ **10.** $x + 2y = -11$ **11.** 5

12. $y = 2x - 14$ **13.** perpendicular **14.** 13 **15.** (2, –8) **16.** yes

CHAPTER 5

1. $6x^2 + 11x - 35$ **2.** $4x^3 - 24x^2$ **3.** $4x^2 + 12x + 9$ **4.** $3(2x - 5)$ **5.** $(w + 7)(w + 7)$

6. $(x + 3)(x - 7)$ **7.** $(2x + 3)(x + 6)$ **8.** $2(5x - 6)(2x + 5)$ **9.** $(2x - 5)(2x + 5)$

10. $(3x + y)(a + 2m)$ **11.** $(2x - 5)(4x^2 + 10x + 25)$ **12.** $\dfrac{5x}{6y}$ **13.** $\dfrac{4}{5}$ **14.** $\dfrac{6x+5}{x+7}$ **15.** $\dfrac{-(5x+2)}{x+6}$

16. $\dfrac{x^2-x-2}{x-5}$ **17.** $x^2 + 4x + 16$ **18.** $4x^2(x + 3)^2$ **19.** 2 **20.** $\dfrac{-1}{2x+7}$ **21.** $\dfrac{x+4}{3x+2}$ **22.** $\dfrac{y-x}{y+x}$

23. $\dfrac{2x+7y}{x+4y}$ **24.** $\dfrac{7}{9}$ **25.** –6 **26.** $\dfrac{Fr}{v^2}$ **27.** $\dfrac{3c}{3-xc}$

CHAPTER 6

1. (3, –1) **2.** (–4, 5) **3.** no solution **4.** 45 μF and 25 μF **5.** 429 ml of 5%, 1071 ml of 12%

6. man 4.4 ft/s, escalator 2 ft/s **7.** $\left(\frac{1}{5}, -\frac{1}{8}\right)$ **8.** (–2, 6, 3) **9.** (3.2, –5.5, 7) **10.** 40 lb pecans, 35 lb raisins, 25 lb candy **11.** 28 **12.** $5m - 24$ **13.** 115 **14.** 21 **15.** T **16.** T **17.** F **18.** F

19. $\dfrac{\begin{vmatrix} 7 & g_1 \\ 10 & g_2 \end{vmatrix}}{\begin{vmatrix} f_1 & g_1 \\ f_2 & g_2 \end{vmatrix}}$ **20.** $\frac{2}{3}$ **21.** $\frac{6}{x-3} + \frac{-2}{x+5}$ **22.** $\frac{2x+3}{x^2+4} + \frac{-8}{x+2} + \frac{7}{x-2}$ **23.** $\frac{4}{2x+5} + \frac{-10}{(2x+5)^2}$

CHAPTER 7

1. $\frac{7}{2}, -3$ **2.** $\pm\frac{3\sqrt{3}}{2}$ **3.** $\pm\frac{7}{4}$ **4.** a) 25 b) $\frac{1}{36}$ **5.** 2, –5 **6.** $\frac{3}{2}, -\frac{5}{3}$ **7.** $-\frac{9}{2}, -2$

8. $\frac{7}{3}, -4$ **9.** $\frac{7}{4}, -\frac{5}{2}$ **10.** $\frac{5}{3}, -\frac{3}{2}$ **11.** w = 18.0 in., l = 48.0 in. **12.** a) 3.00 s or 10.0 s

b) $\frac{13 \pm \sqrt{109}}{2}$, or 11.7 s, 1.28 s **13.** w = 4.20 m, l = 8.60 m **14.** 5.60 cm × 5.60 cm

CHAPTER 8

1. x^{18} **2.** $\frac{1}{x^2}$ **3.** x^3 **4.** $8x^3$ **5.** x^8 **6.** $81x^8w^{12}$ **7.** $\frac{y^7}{x^9}$ **8.** $\frac{w^{12}}{8x^6}$ **9.** 9 **10.** $-\frac{1}{3}$ **11.** 25

12. $10x - \frac{6}{x^2}$ **13.** $4\sqrt{3}$ **14.** $9x\sqrt{2}$ **15.** $\frac{\sqrt{6}}{8}$ **16.** $\frac{4\sqrt{5y}}{5y^2}$ **17.** $3xw\sqrt[3]{5xw^2}$

18. $\frac{\sqrt[3]{45xy^2}}{12xy}$ **19.** $17\sqrt{3}$ **20.** $10x\sqrt{5x}$ **21.** $\frac{39\sqrt{2x}}{14x^2}$ **22.** $15\sqrt{14}$ **23.** 4 **24.** $2\sqrt[4]{9}$

25. $-5 + 2\sqrt{7}$ **26.** $51 - 6\sqrt{30}$ **27.** 61.25 **28.** $x = 5, x \neq -7$ **29.** $\frac{7}{2}$ or $-\frac{4}{3}$ **30.** $\frac{2}{3}$

31. $\pm 3, \pm\dfrac{\sqrt{10}}{2}$ **32.** $\pm\dfrac{1}{5}, \pm\dfrac{1}{2}$ **33.** $\pm 1, 5, 7$

CHAPTER 9

1.

x	-1	0	1	2
y	$\frac{1}{7}$	1	7	49

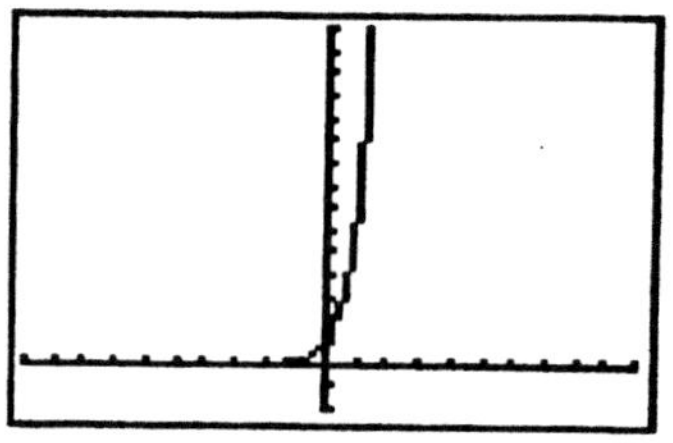

$y = 7^x$

2. 0.0128 **3.** \$83,708.26

4. $\log_3 m = w$ **5.** $c^f = 4$

6.

x	$\frac{1}{3}$	1	3	9
y	-1	0	1	2

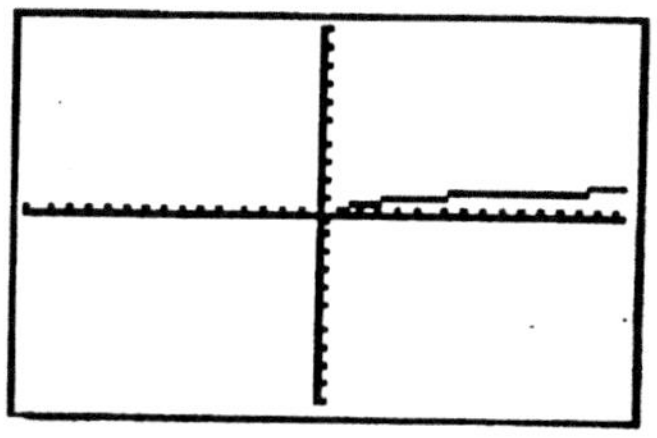

$y = \log_3 x$

7. D **8.** B **9.** A **10.** C

11. $\dfrac{2}{3}\log_5 x + \dfrac{1}{3}\log_5 y - 2\log_5 w$ **12.** $\log_5\left(\dfrac{(x+3)^4}{w^3}\right)$ **13.** 20 **14.** $-x$ **15.** -4 **16.** 2.631

17. -1.186 **18.** 4.595 **19.** 1.063 **20.** -6.372 **21.** 0.7837 **22.** 7.45 **23.** 5510 **24.** 0.00172

25. 1.535 **26.** 6.720 **27.** 15.5 **28.** 0.251 **29.** 2550 **30.** 726 units **31.** 5.113 **32.** 0.973

33. 3.22 **34.** 0.576 **35.** 5.79 **36.** 5.45 **37.** 14.4 yrs **38.** 1.0×10^{-2} w/cm^2

39.

x	1	2	3	4
y	5	20	45	80

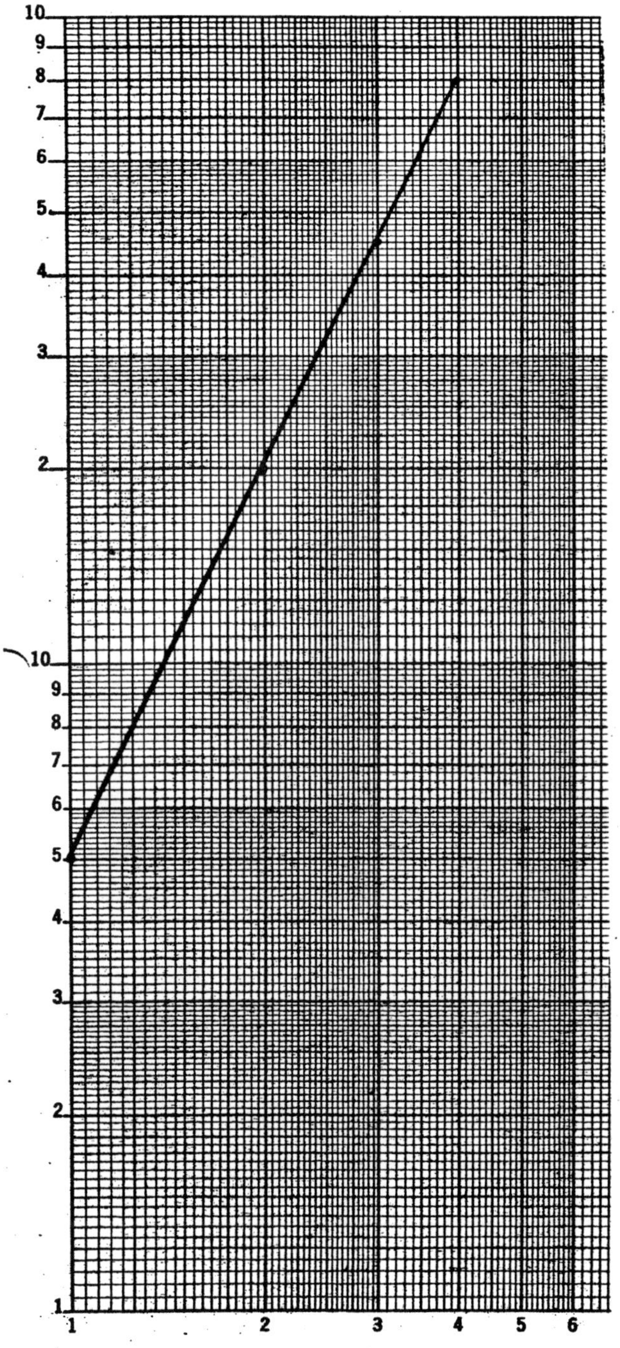

$y = 5x^2$

40.

x	1	2	3	4	5
y	3	9	27	81	243

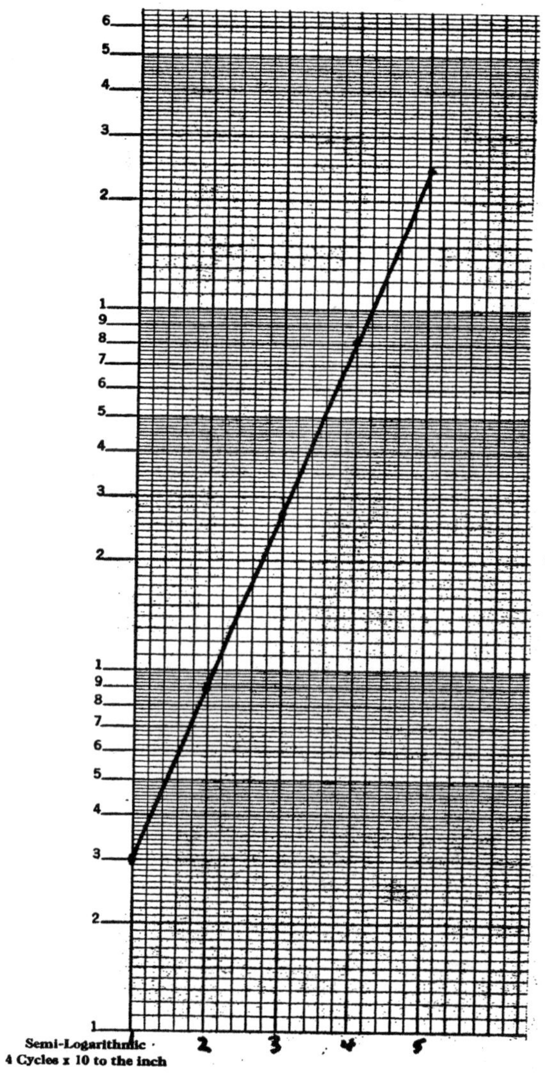

$y = 3^x$

CHAPTER 10

1. 495° **2.** −145° **3.** $\sin\theta = \frac{6}{\sqrt{85}}, \cos\theta = \frac{-7}{\sqrt{85}}, \tan\theta = \frac{-6}{7}$

4. $\csc\theta = \frac{-\sqrt{89}}{8}, \sec\theta = \frac{\sqrt{89}}{5}, \cot\theta = \frac{-8}{5}$ **5.** 55° **6.** 68° **7.** 65° **8.** −0.4195 **9.** −1.381

10. 72.6°, 107.4° **11.** 134.8°, 314.8° **12.** 141.3°, 218.7° **13.** 141.5° **14.** $\frac{\pi}{15}$ **15.** $\frac{-\sqrt{3}}{2}$

16. $\frac{-\sqrt{3}}{3}$ **17.** −0.3857 **18.** 3.85, 5.57 **19.** 144 π radians, 72.0 revolutions **20.** 785 ft/s

CHAPTER 11

1. $B = 24.6°$, $C = 122.8°$, $c = 512$ ft **2.** $A = 113.7°$, $a = 143$ m, $C = 27.6°$

3. $C = 73.8°$, $A = 76.0°$, $a = 27.6$ cm, $C' = 106.2°$, $A' = 43.6°$, $a' = 19.6$ cm

4. not a triangle **5.** $B = 76.4°$, $C = 57.1°$, $c = 41.9$ yd, $B' = 103.6°$, $C' = 29.9°$, $c' = 24.9$ yd

6. $a = 70.6$ cm, $C = 58.3°$, $B = 74.4$ cm **7.** $C = 48.8°$, $A = 72.9°$, $B = 58.3°$

8. $c = 562$ m, $A = 43.6°$, $B = 71.2°$ **9.** 408 ft **10.** 129 ft uphill, 196 ft downhill **11.** $51\bar{0}$ ft

12. 9.0 m at 50.2° **13.** 46.1 ft at 166° **14.** $\mathbf{V}_x = -627$ mi, $\mathbf{V}_y = -606$ mi **15.** 456 m at 308°

16. 109 ft at 254° **17.** $z = 302\ \Omega$, $\phi = 33.1°$ **18.** 317 mph at 63.4°

19. horizontal 27.9 lb, vertical −16.0 lb **20.** $T_1 = T_2 = 575$ lb

CHAPTER 12

1. amp = 4, $P = 2\pi$ **2.** amp = 2, $P = \pi$ **3.** amp $= \frac{1}{2}$, $P = \frac{1}{2}$ **4.** amp = 5, $P = \frac{\pi}{6}$ **5.** amp = 1, $P = 2\pi$, right $\frac{\pi}{4}$ **6.** amp = 2, $P = \frac{2\pi}{3}$, left $\frac{\pi}{24}$ **7.** amp = 4, $P = \frac{\pi}{3}$, right $\frac{\pi}{12}$ **8.** $P = \frac{T}{6}$

9. $P = \pi$ **10.** $P = \frac{\pi}{2}$ **11.** π **12.** see graph **13.** see graph **14.** $y = 2\sin\left(12\pi t + \frac{\pi}{t}\right)$

15. $y = 220 \cos 120\pi t$

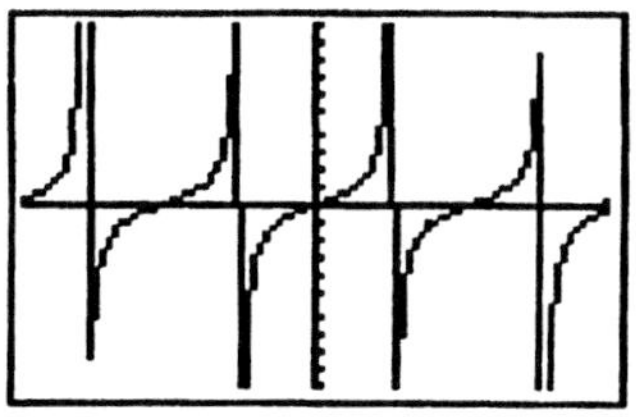

16. $y = \tan 6x$

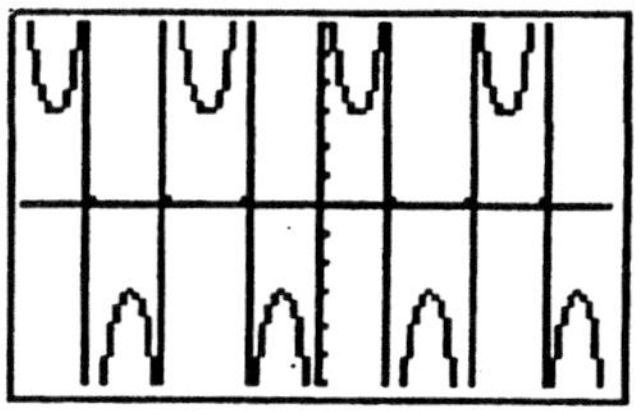

17. $y = 3 \csc 2x$

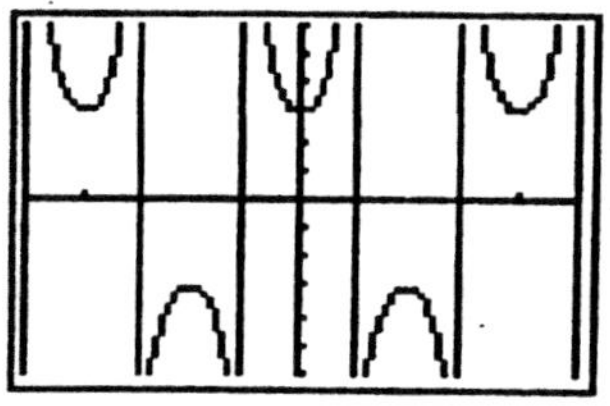

18. $y = 3 \sec 4x$

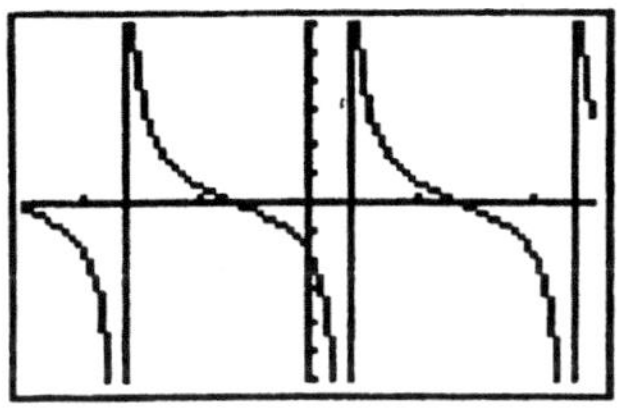

19. $y = \cot\left(x - \frac{\pi}{6}\right)$

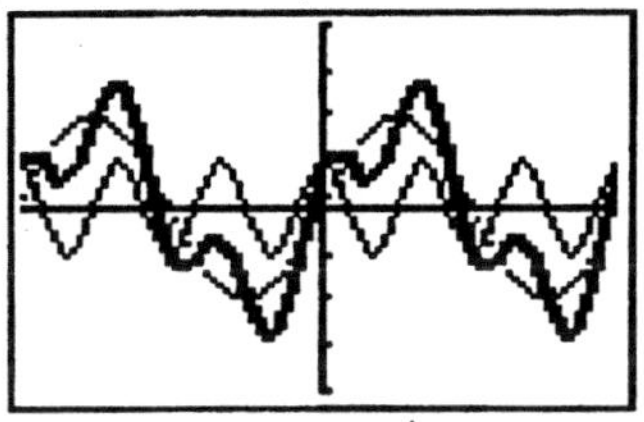

20. $y = 2 \sin x + \cos 3x$

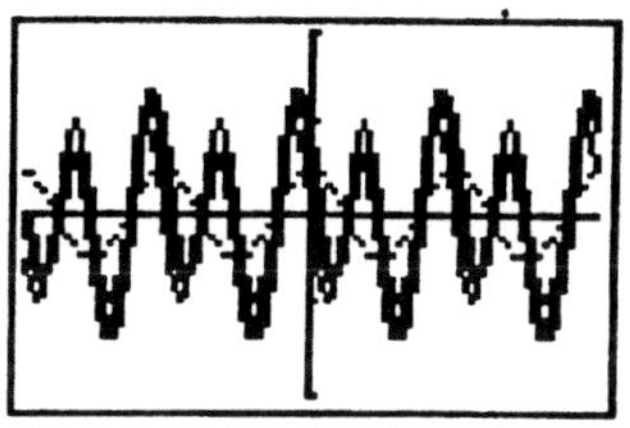

21. $y = \frac{1}{2}\cos 2x - \sin 4x$

CHAPTER 13

1. $\frac{\sin^2 x}{1 - \cos x} = 1 + \cos x$

$$LHS = \frac{\sin^2 x(1+\cos x)}{(1-\cos x)(1+\cos x)}$$
$$= \frac{\sin^2 x(1+\cos x)}{1-\cos^2 x}$$
$$= 1 + \cos x$$
$$= RHS$$

2. $\frac{1+\tan^2 x}{\csc^2 x} = \tan^2 x$

$$LHS = \frac{\sec^2 x}{\csc^2 x}$$
$$= \frac{1}{\cos^2 x} \cdot \frac{\sin^2 x}{1}$$
$$= \tan^2 x$$
$$= RHS$$

3. $\dfrac{\sec x - \cos x}{\tan x} = \sin x$

$$\begin{aligned} LHS &= \frac{\cos x}{\sin x}\left(\frac{1}{\cos x} - \frac{\cos x}{1}\right) \\ &= \frac{1}{\sin x} - \frac{\cos^2 x}{\sin x} \\ &= \frac{1-\cos^2 x}{\sin x} \\ &= \frac{\sin^2 x}{\sin x} \\ &= \sin x \\ &= RHS \end{aligned}$$

4. $\dfrac{\sec^4 x - 1}{\tan^2 x} = \sec^2 x + 1$

$$\begin{aligned} LHS &= \frac{(\sec^2 x + 1)(\sec^2 x - 1)}{\tan^2 x} \\ &= \frac{(\sec^2 x + 1)(\tan^2)x}{\tan^2 x} \\ &= \sec^2 x + 1 \\ &= RHS \end{aligned}$$

5. $\sin\left(x + \dfrac{3\pi}{2}\right) = -\cos x$

$$\begin{aligned} LHS &= \sin x \cos\frac{3\pi}{2} + \cos x \sin\frac{3\pi}{2} \\ &= 0 - \cos x \\ &= RHS \end{aligned}$$

6. $\tan\left(x - \dfrac{\pi}{4}\right) = \dfrac{\tan x - 1}{\tan x + 1}$

$$\begin{aligned} LHS &= \frac{\tan x - \tan\frac{\pi}{4}}{1 + \tan x\left(\tan\frac{\pi}{4}\right)} \\ &= \frac{\tan x - 1}{1 + \tan x} \\ &= RHS \end{aligned}$$

7. $\sin(M - N)$ **8.** $-2\sin M \sin N$ **9.** $\cos 4\theta$ **10.** $25\sin 30t$ **11.** $\dfrac{119}{169}$ **12.** $\dfrac{5}{3}$

13. $\tan\dfrac{\theta}{2} = \dfrac{\sin\theta}{1+\cos\theta}$

$$\begin{aligned} RHS &= \frac{\sin\theta(1-\cos\theta)}{(1+\cos\theta)(1-\cos\theta)} \\ &= \frac{\sin\theta(1-\cos\theta)}{1-\cos^2\theta} \\ &= \frac{1-\cos\theta}{\sin\theta} \\ &= \tan\frac{\theta}{2} \\ &= LHS \end{aligned}$$

14. 105°, 165°, 285°, 345° **15.** 0°, 120°, 240° **16.** 0°, 41.4°, 318.6°

17. $x = 2 + \arcsin(y/5)$ **18.** $\dfrac{5\pi}{6}$

19. $\dfrac{2}{\sqrt{3}}$ **20.** $\sqrt{1-x^2}$

CHAPTER 14

1. -1 **2.** $41+j$ **3.** $\frac{7+11j}{5}$ **4.** $\frac{5\pm\sqrt{23}j}{6}$ **5.** $7\sqrt{2}\,(\cos 315° + j\sin 315°)$ **6.** $\sqrt{7}$ $(\cos 41° + j\sin 41°)$ **7.** $\sqrt{73}\,(\cos 111° + j\sin 111°)$ **8.** $\frac{-5\sqrt{3}}{2}-\frac{5}{2}j$ **9.** $-2.73+3.25j$

10. $3.38-7.25j$ **11.** $5e^{5.64j}$ **12.** $8\sqrt{2}e^{3.93j}$ **13.** $2\sqrt{34}e^{2.11j}$ **14.** $1.26+3.80j$

15. $-4.92-3.43j$ **16.** $4.11-5.67j$ **17.** $21\,e^{3.4j}$, $-20.3-5.37j$ **18.** $10e^{7.35j}$, $4.83+8.76j$

19. $4\,e^{8.05}$, $-0.779+3.92j$ **20.** $42(\cos 91° + j\sin 91°)$ **21.** $20(\cos 102° + j\sin 102°)$

22. $4\,e^{1.8j}$, $-0.909+3.90j$ **23.** $7\,e^{-4.55j}$, $-1.13+6.91j$ **24.** $0.429\,e^{-7.3j}$, $0.226-0.365j$

25. $16(\cos 184° + j\sin 184°)$ **26.** $0.8(\cos 115° + j\sin 115°)$ **27.** $-62.5+108j$

28. $172-172j$ **29.** $29.4+840j$ **30.** $2^{1/8}(\cos 78.8° + j\sin 78.8°)$,

$2^{1/8}(\cos 169° + j\sin 169°)$, $2^{1/8}(\cos 259° + j\sin 259°)$, $2^{1/8}(\cos 349° + j\sin 349°)$

CHAPTER 15

1. 2×4 **2.** $x=-5, y=2$ **3.** $\begin{bmatrix}-2 & -11\\ 3 & 10\end{bmatrix}$ **4.** $\begin{bmatrix}9 & -30\\ -3 & 21\end{bmatrix}$ **5.** not defined **6.** $\begin{bmatrix}4 & -7 & 36\\ 32 & 35 & -50\end{bmatrix}$

7. $[19]$ **8.** $\begin{bmatrix}-3 & 56 & 63\\ 8 & 27 & 33\\ 62 & -56 & -18\end{bmatrix}$ **9.** $\begin{bmatrix}1 & -2\\ -0.5 & 1.5\end{bmatrix}$ **10.** $\begin{bmatrix}-3 & -8 & 0\\ -2 & -5 & 0\\ 0.5 & 1.25 & 0.25\end{bmatrix}$

11. $x=3, y=-5, z=7$ **12.** $x=42, y=38, z=20$

CHAPTER 16

1. 9 **2.** 7 **3.** x: 5, 2/3 repeated; y: −20 **4.** x: −3, 2 repeated; y: 12

5. $f(x) = 6x^3 + 13x^2 - 78x + 35$ **6.** 7 with multiplicity 3 and −4 with multiplicity 2

7. even **8.** −2.949, 0.7828, 2.166 **9.** −0.6383, 0.6128, 8.185 **10.** 8, −5, $3 \pm j$

11. $(x+7)^3(x^2 - 8x + 41) = 0$ or $x^5 + 13x^4 + 20x^3 + 28x^2 + 3283x + 14063 = 0$

12. $\pm 3j$, 2, −1

CHAPTER 17

1. $x > -2$ **2.** $-5 \le x < 1$ **3.** $x < -3$ or $x \ge 7$ **4.** $x > -8$ **5.** $x \le -10$ **6.** $x > -75$ **7.** 10, −3

8. $-11 \le x \le 5$,

9. $x > 12$ or $x < 2$,
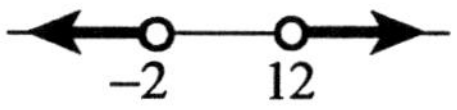

10. $-4 \le x \le 9$,

11. $-7 < x < 2.5$ **12.** $x \le -6$ or $x > 5$

13. $x < 0$ or $x \ge 1$ **14.** $y \ge 2x - 7$ **15.** $y < 4$ **16.** $4x - 8y \ge 15$

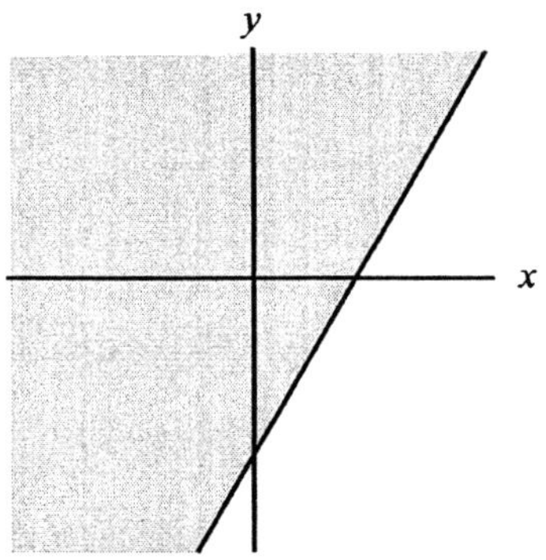

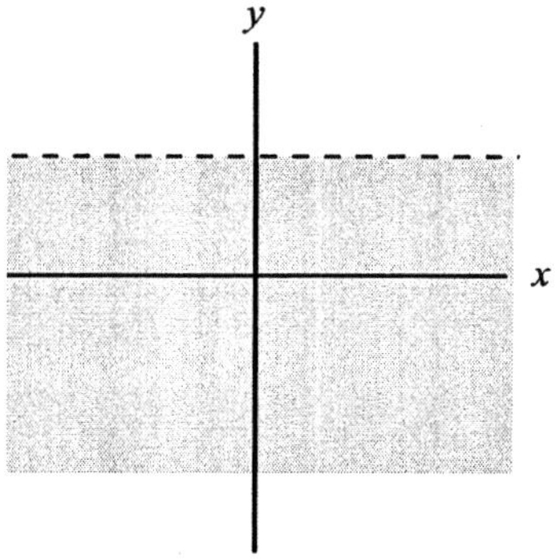

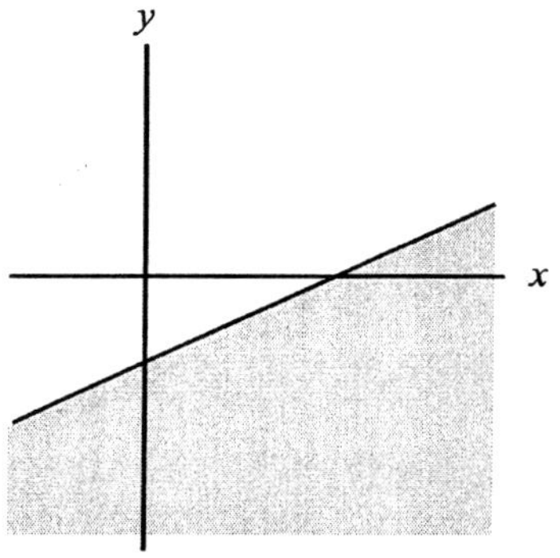

17. $y \ge x^2 - 4$
$y \le 2$

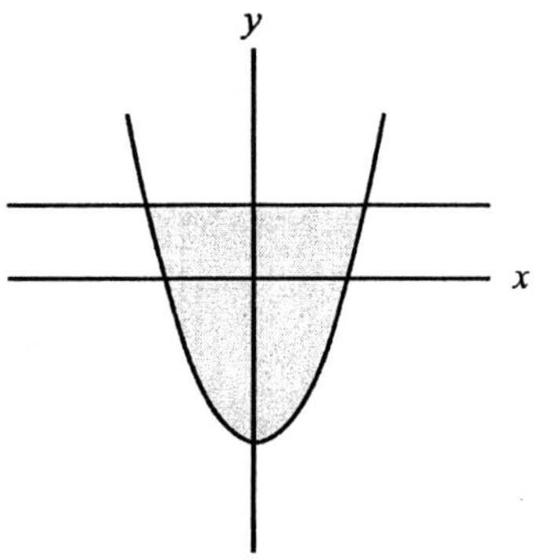

18. $y \le 2x + 5$
$y \ge 0$
$x \ge 0$

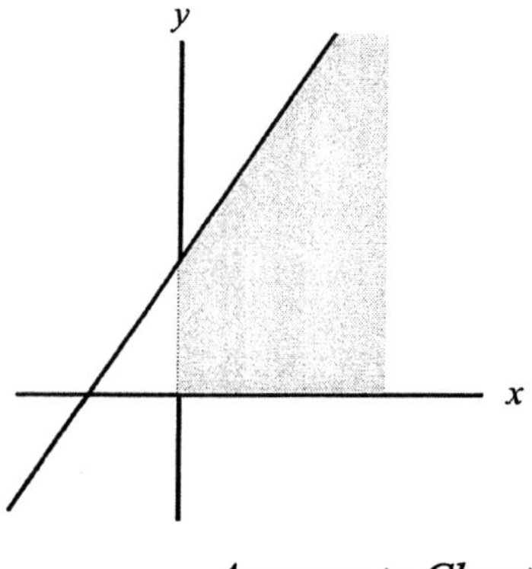

CHAPTER 18

1. 23 **2.** −19 **3.** 23 **4.** −38 **5.** 78 **6.** −21 **7.** −476 **8.** \$39,700 **9.** 324 **10.** -2^{-11} **11.** -3^{-5}

12. −2560 **13.** $\frac{45}{8}$ **14.** 728 **15.** $\frac{364}{3}$ **16.** $91\frac{1}{8}$ °F **17.** $\frac{75}{4}$ **18.** $\frac{96}{5}$ **19.** not possible

20. $x^4 + 16x^3y + 96x^2y^2 + 256xy^3 + 256y^4$

21. $64x^6 - 960x^5 + 6000x^4 - 20{,}000x^3 + 37{,}500x^2 - 37{,}500x + 15{,}625$

22. $1024x^5 - 3840x^4y + 5760x^3y^2 - 4320x^2y^3 + 1620xy^4 - 243y^5$ **23.** $-175{,}000x^4y^3$

24. $30{,}375x^4y^2$

CHAPTER 19

1.

Bushels	148-151	152-155	156-159	160-163	164-167	168-171	172-175	176-179	180-183
Freq.	5	3	3	7	3	6	6	2	5

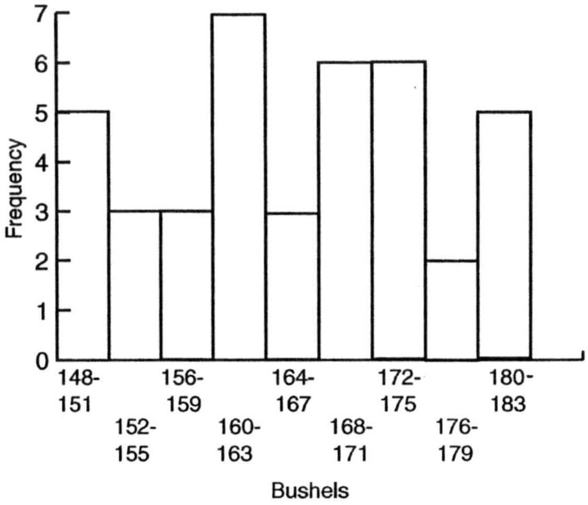

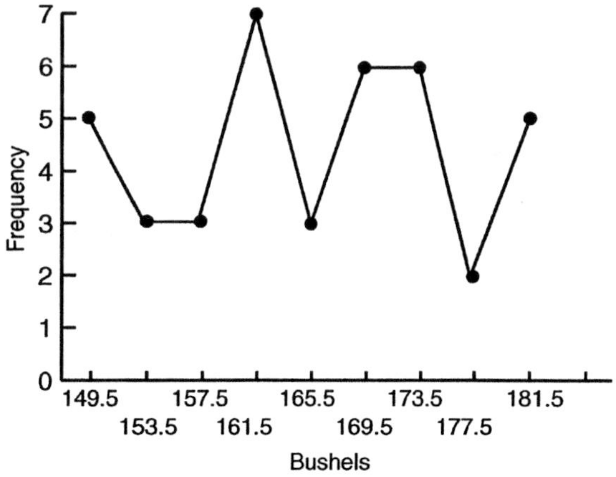

2. $\bar{x} = 27.9$, med $= 28.5$, mode $= 24$ **3.** $\bar{x} = 26.1$, med $= 26$, mode $= 27$ **4.** $r = 8$, $s_x = 2.67$

5. $r = 9$, $s_x = 3.51$ **6.** 69.9% yes, 94.4% yes **7.** $y = -2.5x + 3.1$ **8.** $y = -0.89x^2 - 3.3x + 6.8$

9. $y = 2.0(7.4)^x$ **10.** The process is capable.

CHAPTER 20

1.

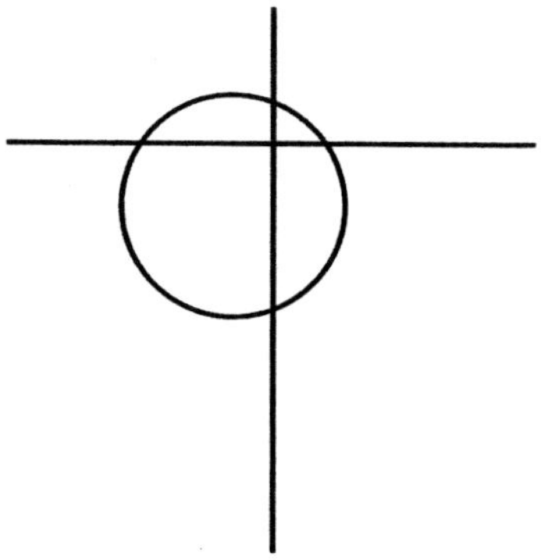

$(x+2)^2 + (y+3)^2 = 16$

2. $(x-4)^2 + (y+1)^2 = 3^2$

3. $(x-7)^2 + (y-3)^2 = 26$

4. $C(3, -2)$, $r = 4$

5. $F(-5, 0)$, directrix $x = 5$

6. $x^2 = -12y$ **7.** $x^2 = 24y$

8. $V\left(\frac{7}{4}, \frac{-169}{8}\right)$, axis: $x = \frac{7}{4}$

9. $V(0, \pm 8)$, $F(0, \pm\sqrt{55})$, 16, 6

10. $V(\pm 12, 0)$, $F(\pm\sqrt{119}, 0)$, 24, 10 **11.** $\frac{x^2}{49} + \frac{y^2}{9} = 1$ **12.** $\frac{x^2}{25} + \frac{y^2}{64} = 1$

13. $V(\pm 6, 0)$, $F(\pm 10, 0)$, T:12, C:16, $y = \pm\frac{4}{3}x$ **14.** $V(0, \pm 10)$, $F(0, \pm\sqrt{181})$, T:20, C:18, $y = \pm\frac{10}{9}x$ **15.** $\frac{x^2}{81} - \frac{y^2}{63} = 1$ **16.** $\frac{y^2}{25} - \frac{x^2}{24} = 1$ **17.** $8(y-4) = (x-3)^2$

18. $\frac{(x+3)^2}{25} + \frac{(y-7)^2}{169} = 1$ **19.** $\frac{(x+1)^2}{9} - \frac{(y-2)^2}{4} = 1$, hyperbola, $C(-1, 2)$ **20.** circle

21. $(4.5, \pm 2.6)$ **22.** $(-4, 225\text{E})$, $(-4, -135\text{E})$, $(4, -315\text{E})$

23.

θ	0	$\frac{\pi}{4}$	$\frac{\pi}{2}$	$\frac{3\pi}{4}$	π	$\frac{5\pi}{4}$	$\frac{3\pi}{2}$	$\frac{7\pi}{4}$	2π
r	6	1.8	0	1.8	0	10	12	10	6

24. $(2.5,\ 2.5\sqrt{3})$

25. $(8,\ 1.25\ \pi)$

26. $r = 7$

27. $y = \frac{1}{10}(x^2 - 25)$

28. $r = 7\sin\theta$

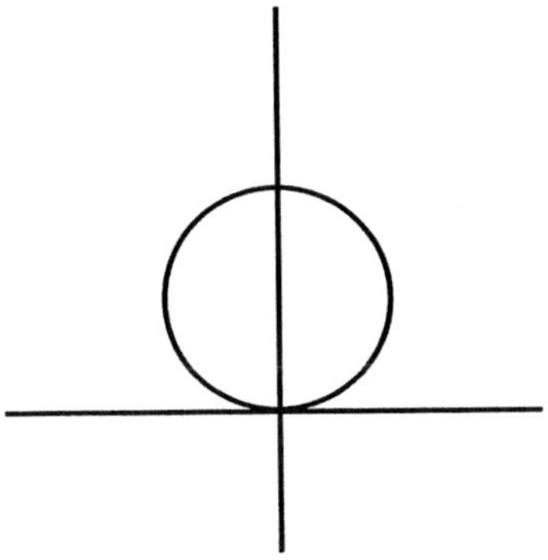

29. $r = 8 + 4\cos\theta$

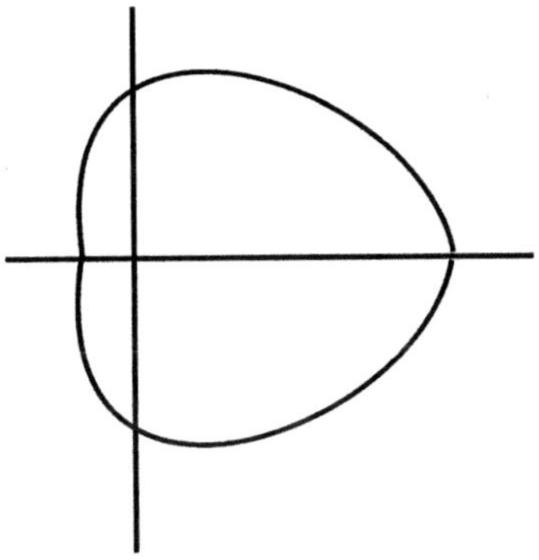

30. $r = 5 + 12\sin\theta$

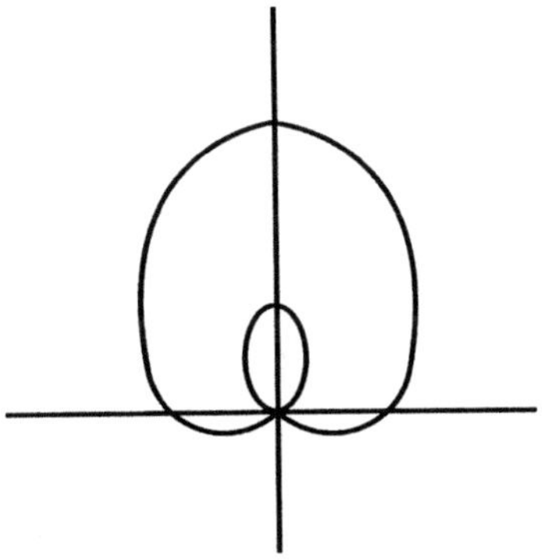

31. $r = 3\sin 5\theta$

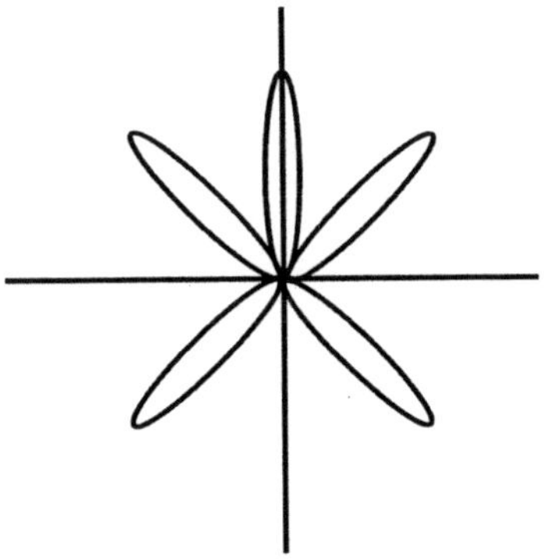

32. $r = 6\cos 4\theta$

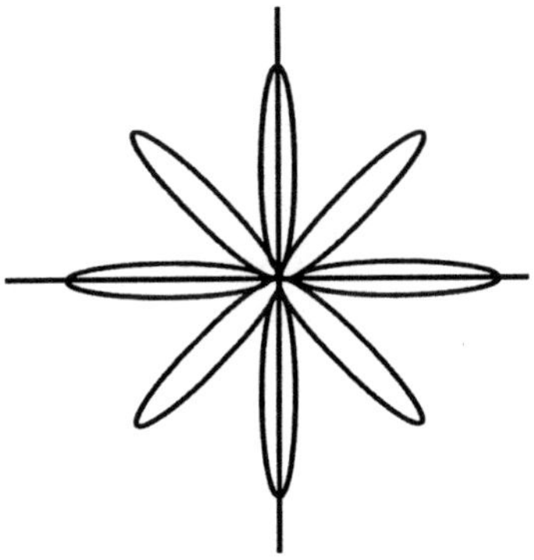

CHAPTER 21

1. $f(4) = 126$ **2.** $f(5) = \sqrt{5}$ **3.** $\Delta s = 5\Delta t,\ \bar{v} = 5$ **4.** $\Delta s = 6t\Delta t + 3(\Delta t)^2 = 10\Delta t,\ \ \bar{v} = 6t - 10 + 3\Delta t$

5. $\bar{v} = 44$ m/s **6.** $\bar{v} = 24$ m/s **7.** 10 **8.** $\frac{1}{2}$ **9.** 5 **10.** $-\frac{15}{2}$ **11.** 0 **12.** 8

13. 6 **14.** $y = 24x - 57$ **15.** (0, 0) and $\left(\frac{5}{3}, -\frac{125}{27}\right)$ **16.** $y' = 4x - 5$ **17.** $y' = \frac{-2}{(2x+7)^2}$

18. $y = 3x + 5$ **19.** $y' = 14x^{13}$ **20.** $y' = 8x^3 - 6x$ **21.** $y' = 4\sqrt{6}x^3$ **22.** $y' = \frac{-16}{x^9}$

23. $y' = \frac{-2.5}{x^3\sqrt{x}}$ **24.** $f'(2) = 92$ **25.** $y' = 7x^6 - 12x^5 + 25x^4 - 40x^3$

26. $y' = 30x^9 + 45x^8 - 56x^7$ **27.** $y' = \frac{-9}{(4x-9)^2}$ **28.** $y' = \frac{-6x^2 - 18x - 10}{(3x^2 - 5)^2}$

29. $y' = 27(3x - 7)^8$ **30.** $y' = -30x^2(x^3 - 8)^4$ **31.** $y' = \frac{24(3x+2)(4x-1)^5 - 12(4x-1)^6}{(3x+2)^5}$

32. $y' = -\frac{8}{3}$ **33.** $y' = \frac{-9x}{4y}$ **34.** $y' = \frac{3x^2 - 2}{(1+9x^2)}$ **35.** $y' = \frac{7y + 3x^2(x-3y)^2}{7x}$

36. $y = \frac{6}{5}x - 4$ **37.** $y' = 20x^4 - 12x^3 + 16x,\ y'' = 80x^3 - 36x^2 + 16,\ y''' = 240x^2 - 72x$

38. $y' = \frac{2}{2y+3},\ y'' = \frac{-8}{(2y+3)^3}$

CHAPTER 22

	Intercepts	Intervals $y \geq 0$	Symmetry	Asymptotes
1.	none	none	about x-axis	$x = 3$
2.	(2, 0), (–3, 0), (6, 0), (0, 36)	$[-3, 2] \cup [6, \infty)$	none	none
3.	$\left(0, -\frac{4}{3}\right)$	$\left(\frac{3}{5}, \infty\right)$	none	$x = \frac{3}{5}$, x-axis
4.	(0, 0)	$(-4, 0] \cup (5, \infty)$	none	$x = -4$ or $x = 5$

5. at $x = 1$ rel min = 0, at $x = \frac{-7}{3}$ rel max $= \frac{200}{9}$ **6.** at $x = \pm\sqrt{3}$ rel min = 4, at $x = 0$ rel max = 5 **7.** a) increasing for $x < -3$, $-3 < x < 0$; decreasing for $0 < x < 3$, $x > 3$ b) rel max at $\left(0, \frac{-2}{3}\right)$ c) critical points ± 3, $2 \pm \sqrt{13}$, concave upward for $x < -3$, $x > 3$, concave downward for $-3 < x < 3$ d) no point of inflection

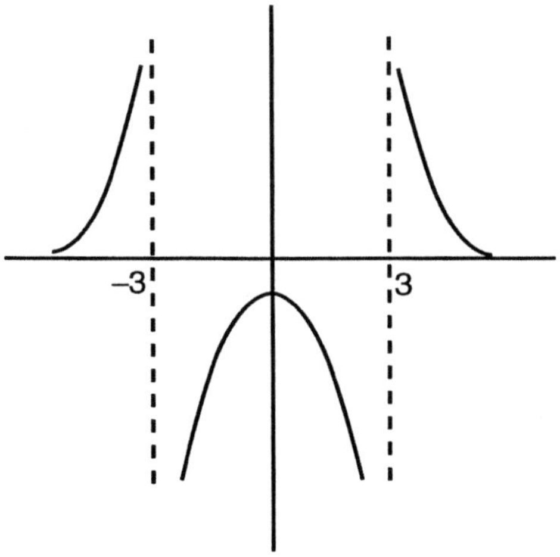

8. 2.64575 and –2.64575

9. 9. 0.321751 and 3.46334

10. 30 cm × 30 cm × 15 cm **11.** 9 Ω

12. $\frac{dy}{dt} = 180$ **13.** 2.7 in./h **14.** 1.56 cm/s

15. $dy = \frac{-9}{25} x\,dx$ **16.** $dV = 250$

17. $\frac{dV}{V} = 1.2\%$

18. $dV = 45\pi m^3 = 141 m^3$; 1,099,800 kg

CHAPTER 23

1. $y' = -15 \sin 15x$ **2.** $y' = 6 \cos(2x + 1)$ **3.** $y' = \frac{6x \sin 2x - 2(3x^2 + 5)\cos 2x}{\sin^2 2x}$

4. $y' = -8 \sin 6x \sin 8x + 6 \cos 8x \cos 6x$ **5.** $y' = \dfrac{(x+2)\sec^2\sqrt{x^2+4x}}{\sqrt{x^2+4x}}$

6. $y' = \dfrac{x \sec x \tan x - 2\sec x}{x^3}$ **7.** $y' = -2 \sin 3x \cot x \csc^2 x + 3 \cos 3x \cot^2 x$

8. $y' = -3 \csc x \cot x(1 + \csc x)^2$ **9.** $y' = \dfrac{5}{\sqrt{2x - x^2}}$ **10.** $y' = \dfrac{-12}{4+x^2}$ **11.** $y' = \dfrac{8x}{\sqrt{2x^2 - x^4}}$

12. $y' = \dfrac{-8\text{Arccsc}^3 2x}{|2x|\sqrt{4x^2-1}}$ **13.** $y' = \dfrac{4\log e}{4x+1}$ **14.** $y' = \dfrac{2x}{x^2-3}$

15. $\dfrac{dy}{dx} = \dfrac{(x-3)^2(5x+4)}{(2x-7)(6-x)}\left[\dfrac{2}{x-3} + \dfrac{5}{5x-4} - \dfrac{2}{2x-7} + \dfrac{1}{6-x}\right]$

16. $\dfrac{dy}{dx} = -x \tan x(\cos x)^x (\cos x)^x \ln(\cos x)$ **17.** $y' = (3x^2+3)e^{x^3+3x}$

18. $y' = 2e^{2x} \cos e^{2x}$ **19.** $y' = e^x(x^2 + 2x)$ **20.** $\dfrac{1}{2}$ **21.** -24

22. rel max at $(-e^{-1}, 2e^{-1})$, rel min at $(e^{-1}, -2e^{-1})$

CHAPTER 24

1. $\dfrac{x^{24}}{8} + C$ **2.** $\dfrac{2}{9}(3x+1)^{3/2} + C$ **3.** $-\dfrac{1}{24}(4x^3-5)^{-2} + C$ **4.** $\dfrac{4}{45}(3x^5+10x^3)^{3/4} + C$

5. $y = \dfrac{1}{8}(x^2-6)^4 + 3$ **6.** a) $s = -16t^2 + 44t + 28$ b) 3.28 seconds **7.** $V_c = 1020\ V$

8. $\dfrac{203}{6}$ **9.** $\dfrac{49}{3}$ **10.** $54\dfrac{3}{4}$ **11.** 29.5 **12.** $\dfrac{122}{3}$ **13.** 16 **14.** $-144\dfrac{3}{4}$

CHAPTER 25

1. $114\dfrac{1}{3}$ **2.** 2 **3.** 36π **4.** $\dfrac{81\pi}{3}$ **5.** $\dfrac{27\pi}{2}$ **6.** $\left(20\dfrac{1}{3}, 0\right)$ **7.** $\left(-\dfrac{7}{11}, \dfrac{35}{11}\right)$ **8.** $\dfrac{408}{35}$ cm from lighter end **9.** (0, 3.6) **10.** $I_y = 129.6$, $m = 72$, $R = 1.34$ **11.** $I_x = 72{,}817\pi$, $m = 1024\pi$, $R = 8.43$ **12.** 53,913.6 lb rounded to 53,900 lb

CHAPTER 26

1. $\frac{1}{9}(x^3+2x^2)^9 + C$ **2.** $\frac{1}{6}e^{6x} + C$ **3.** $\frac{1}{10}\sin^2 5x + C$ **4.** $\frac{1}{6}\ln\left|3x^2+6x\right| + C$

5. $e^{x^3} + C$ **6.** 2 ln 7 or ln 49 **7.** $\frac{1}{5}\tan(5x+10) + C$ **8.** $\frac{1}{3}\sec 3x + C$

9. $\frac{1}{25}\cos^5 5x - \frac{1}{15}\cos^3 5x + C$ **10.** $\frac{1}{16}\tan^2 8x + \frac{1}{8}\ln\left|\cos 8x\right| + C$ **11.** $\frac{1}{3\sqrt{5}}$ Arc sin $3x + C$

12. $\frac{1}{10}\text{Arc}\tan\frac{5x}{2} + C$ **13.** $\frac{5}{x+2} + \frac{3}{x-7}$, $\ln\left|(x+2)^5(x-7)^3\right| + C$ **14.** $\frac{3}{x} + \frac{8}{x-1} - \frac{4}{(x-1)^2}$,

$\ln\left|x^3(x-1)^8\right| + \frac{4}{x-1} + C$ **15.** $\frac{x}{10}e^{10x} - \frac{1}{100}e^{10x} + C$ **16.** $\frac{1}{2}x^2\ln|x| - \frac{1}{4}x^2 + C$

17. $\frac{x}{49\sqrt{49-x^2}} + C$ **18.** $\frac{(x^2-50)}{3}\sqrt{x^2+25}$ **19.** #5 from Table 6: $\frac{1}{12}\ln\left|\frac{4x}{3+4x}\right| + C$

20. #59 from Table 6: $\frac{e^x(2\sin 5x - 5\cos 5x)}{29} + C$ **21.** 0.470006 **22.** 8.696 23. $\frac{1}{4}$ 24. 6

CHAPTER 27

1.

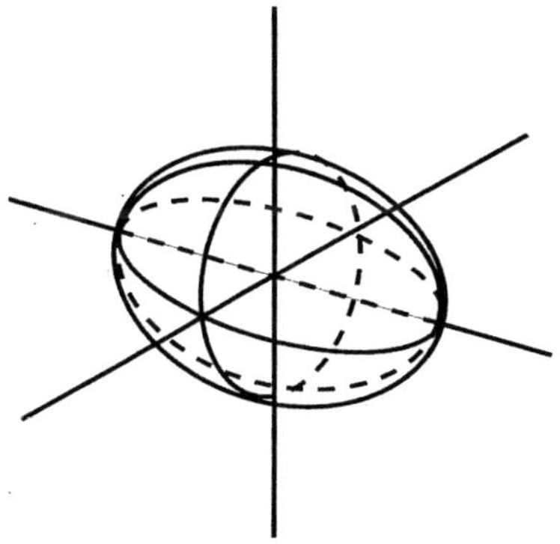

2. $7\sqrt{5}$

3. $\frac{\partial z}{\partial x} = e^{4x}(4\cos xy - y\sin xy)$, $\frac{\partial z}{\partial y} = -xe^{4x}\sin xy$

4. −72

5. $dz = (8x+2y)\,dx + (2x+9y^2)dy$

$$9x^2 + y^2 + 4z^2 = 36$$

Ellipsoid

6. 14.2 Ω **7.** (2.5, 2, 7.5), $D(2.5, 2) = -24$, saddlepoint **8.** $-254.9\overline{3}$ **9.** $V = 16\frac{2}{3}$

CHAPTER 28

1. converges **2.** converges **3.** $r = \frac{1}{5}$, converges **4.** diverges by integral test **5.** converges absolutely **6.** diverges **7.** converges conditionally **8.** $-\infty < x < \infty$ **9.** $-4 < x \le 6$

10. $f(x) = 1 + 2x + 2x^2 + \frac{4}{3}x^3 + \Lambda$ **11.** $f(x) = 3x - \frac{9}{2}x^3 + \frac{81}{40}x^5 + \Lambda$

12. $f(x) = 1 + \frac{x}{2!} + \frac{x^2}{3!} + \frac{x^3}{4!} + \Lambda$ **13.** 0.7639 **14.** $f(x) = \frac{1}{4} - \frac{3}{8}(x-2) + \frac{3}{8}(x-2)^2 - \frac{5}{16}(x-2)^3 + \Lambda$

15. $f(x) = \frac{\sqrt{2}}{2} - \frac{\sqrt{2}}{2}\left(x - \frac{\pi}{4}\right) - \frac{\sqrt{2}}{4}\left(x - \frac{\pi}{4}\right)^2 + \frac{\sqrt{2}}{12}\left(x - \frac{\pi}{4}\right)^3 + \Lambda$ **16.** 3.3199

17. $f(x) = 3\pi - 6\sin x - 3\sin 2x - 2\sin 3x - \frac{3}{2}\sin 4x + \Lambda$

CHAPTER 29

1. third order, second degree **2.** $\frac{x}{4}(5x^4 + 15) - \frac{1}{4}(x^5 + 15x) = x^5$ **3.** $\ln\left[(y-1)\sqrt{x^2+3}\right] = C$

4. $\frac{1}{y} = \frac{1}{2x^2} + \frac{3}{32}$ **5.** $\text{Arctan}\frac{x}{y} = \frac{1}{3}y^3 + C$ **6.** $xy = x^3 + Cx^{-1}$ **7.** $y = xy^2 - 33x$

8. $y = -e^{-3x^2}\cos x + Ce^{-3x^2}$ **9.** $y = \frac{5}{6}x^2 + Cx^{-4}$ **10.** 95 years **11.** $i = 1.75 + 3.25e^{-600t}$

12. 276EF

CHAPTER 30

1. $y = k_1e^{3x} + k_2e^{-7x}$ **2.** $y = k_1 + k_2e^{-8x}$ **3.** $y = -2e^{6x} + 3e^{-2x}$ **4.** $y = k_1e^{11x} + k_2xe^{11x}$

5. $y = e^{2x}(k_1 \sin 5x + k_2 \cos 5x)$ **6.** $y = 7\sin 9x + 3\cos 9x$

7. $y = k_1e^{3x} + k_2e^{-5x} - \frac{1}{15}x - \frac{2}{225} - \frac{1}{7}e^{2x}$ **8.** $y = k_1 \sin x + k_2 \cos x - 2x \cos x$

9. $x = \frac{1}{3}\cos 8t$ **10.** $x = C_1e^{-5.24t} + C_2e^{-9.16t}$ overdamped **11.** $i = k_1e^{-8.11t} + k_2e^{-617t}$

12. $\frac{s-4}{(s-4)^2+64} + \frac{5}{(s-10)^2}$ **13.** $\frac{1}{s^5} + \frac{36}{s(s^2+36)}$ **14.** $\sin 9t + \frac{t^2e^{3t}}{2}$

15. $y = 7e^{8t} + 10e^{-3t}$ **16.** $y = \frac{t^3e^{4t}}{6}$